T0109496

THE
LEGACY
OF LUNA

THE
LEGACY
OF LUNA

THE STORY OF A TREE,
A WOMAN, AND
THE STRUGGLE TO
SAVE THE REDWOODS

JULIA BUTTERFLY HILL

HarperOne
An Imprint of HarperCollins*Publishers*

HarperOne

Grateful acknowledgment is given to the following individuals and organizations for the photographs that appear in this book. Julia in tree with rope, Julia on top of tree, Julia in tree on phone, © Thomas de Soto; Julia on platform, © Charlie Neuman, reprinted courtesy of the *San Diego Union-Tribune;* Julia looking up, © R. Pfotenhauer; Aerial view of clearcutting © Shaun Walker; Julia on the ground © Shaun Walker/OtterMedia.com. All other photographs are from the author's private collection. All artwork and poetry that appears in the book © Julia Butterfly Hill.

This book is printed on paper made from 100% post-consumer recycled fibers and is processed in a totally chlorine free process using soy-based ink. All of the author's profits will go to the Circle of Life Foundation.

HarperCollins books may be purchased for educational, business, or sales promotional use. For information, please e-mail the Special Markets Department at SPsales@harpercollins.com.

HarperCollins Web site: http://www.harpercollins.com
HarperCollins®, ✦®, and HarperOne™ are
trademarks of HarperCollins Publishers.

FIRST HARPERCOLLINS PAPERBACK EDITION PUBLISHED IN 2001

Designed by Joseph Rutt

Library of Congress Cataloging-in-Publication Data
Hill, Julia Butterfly.
The legacy of Luna : the story of a tree, a woman, and the struggle to save the
redwoods / Julia Butterfly Hill.
p. cm.
ISBN 978-0-06-251659-6
1. Hill, Julia Butterfly. 2. Women conservationists—California—Humboldt
County—Biography. 3. Luna (Calif. : Tree) 4. Old growth forest conservation—
California—Humboldt County. 5. Logging—California—Humboldt County.
6. Pacific Lumber Company. I. Title.
SD129.H53 A3 2000
333.75'16'092—dc21
[B]
99-088633

23 24 25 26 27 LBC 31 30 29 28 27

This book is dedicated to the essence of Luna:
strength, endurance, commitment,
and love.

Also, to the magic of the Earth under our feet,
and the power of individuals committed
to living their truth.

*These kings of the forest, the noblest of a noble race, rightly
belong to the world, but as they are in California, we cannot
escape the responsibility as their guardians. Fortunately, the
American people are equal to this trust.*

John Muir

CONTENTS

THE STAFFORD
STORY

Mike O'Neal woke early in the morning of December 31, 1996, to hushed silence. "Thank goodness the rain has stopped," he thought.

California's Humboldt County, located in the northernmost part of the state, traditionally gets a lot of precipitation. Locals joke that the area has three seasons: July, August, and rain. Even by those standards, however, the year's deluge had been impressive. So the night before, Mike, a mustached, well-fed, mobile-home mover and single parent, had made sure to check the creek that runs below his house in the tiny town of Stafford. Any debris washed down from the mountain directly behind clogs up the culvert, which backs up the creek and washes out his foundation.

Suddenly, Mike heard a series of sharp, snapping cracks that shattered the morning stillness. He could tell they were coming from up the hill. He ran to his nine-year-old daughter's bedroom window on the second floor of his house, only to see huge redwood and Douglas fir trees breaking off one after another and slamming down to the ground. In an instant Mike realized what was happening: the mountain above his house had been loosened by the heavy rains and was now sliding down, breaking off all the trees in its path.

Mike ran out of the house, checking the culvert. It was now bone dry. That could mean only one thing: something was damming the water—and that something was a giant wall of trees, rocks, and stumps the width of a football field and twenty feet high. And it was headed straight for his home.

Mike spun around and ran to rouse his sleeping daughter and neighbors. From the looks of things, the flow was going to hit his neighbor's house first, then his, before reaching the homes closer to the freeway.

"The mountain's sliding," he yelled to Kim and Jennie Rollins, who lived next door with their son, Russell (then nine years old), and mother-in-law, Viola Withlow. "What the hell?" asked Kim, still half asleep, struggling to get one leg into his pants when he opened the door. "Look!" Mike responded, pointing up the hill.

Kim whirled back to the house to get his family out. Mike took off in the other direction, grabbed his daughter, then banged on the door of the mobile home parked in front of his house. Assuming those families farther away would be able to recognize the danger in time, Mike and his neighbors bundled into their cars and fled.

The massive, lavalike flow bounced off the corner of Mike's house, then hit the Rollins's home dead-on, lifting it off its foundation and filling it with stumps, trees, and rocks before steamrolling on toward the remaining six houses to the north and hurling a pickup truck into one of them.

Miraculously, no one was hurt by the slide, which would fill the entire basin where their homes were located with eight to seventeen feet of mud and debris. Those dispossessed by the slide spent New Year's—and the subsequent weeks and months—in friends' or relatives' houses or in motels. One unlucky soul, with no better option, was forced to live in his car. Some residents, unable to bear the sense of impending doom, relocated altogether. Those unable to afford rent or hotel charges on top of their mortgage payments stayed. They had no choice.

But with a newly liquefied mountain teetering above, even those Stafford residents whose homes had been spared were deeply worried. They knew that since one valley was filled, any additional surge of water would mean that more mud and debris would roll off sideways, possibly taking out their homes. A wall of mud hung over the town, just waiting for the next rain to bring it down.

Stafford residents blamed the slide on the fact that the steep slope above them had been clear-cut by the Pacific Lumber Company. These were people who lived in logging country. They knew that when a hillside was stripped of trees, nothing was left to hold the dirt and rocks in the heavy rains. Still, people were reluctant to sue Pacific Lumber. It had been pretty much the only business in the area for a hundred fifty years. Almost everyone had an intricate family tie to the company—a grandfather worked in the sawmill, a brother in the woods, an uncle on a logging truck. Challenging the hand that fed them was difficult at best.

"To say Pacific Lumber was bad is just as close as you could get to saying that God was bad. Nobody goes up against God," said a frustrated Mike O'Neal, trying to rustle up support for a lawsuit.

The fact that "God" owned Scotia—America's only remaining company town, lying immediately to the north of Stafford—only made matters worse.

Three days after the slide, Governor Pete Wilson flew over Stafford. After viewing the devastation firsthand, he appointed a professor of civil and environmental engineering from the University of Michigan named Donald H. Gray to investigate the cause of the mud slide.

Pacific Lumber hoped that Professor Gray would agree with its position that the slide had been an act of God, a manifestation of a natural process precipitated by December's intense storms. Professor Gray disagreed, however, and in a letter dated January 8, 1997, he detailed the role vegetation plays in dispersing and absorbing excess moisture as well as the adverse effects of logging roads, skid trails, and landings on the mountain's structural stability. "Both surface and subsurface flow can be diverted, concentrated, and otherwise modified in ways that can have profound consequences," he wrote, pointing to the many studies that "have shown that timber harvest sites have a disproportionately high incidence of slope failures compared to natural, undisturbed areas."

Or, in Mike O'Neal's significantly less scientific words: "Act of God, my ass. This was clear-cutting. Only clear-cutting could have caused this."

Immediately following Professor Gray's opinion, the California Department of Forestry approved a plan to clear-cut the slope directly next to the slide where a tree—soon to be named Luna—and a woman named Julia Butterfly Hill would change the environmental movement forever.

THE
LEGACY
OF LUNA

ONE

FIGHTING FEAR
WITH A FORK

Fierce winds ripped huge branches off the thousand-year-old redwood, sending them crashing to the ground two hundred feet below. The upper platform, where I lived, rested in branches about one hundred eighty feet in the air, twenty feet below the very top of the tree, and it was completely exposed to the storm. There was no ridge to shelter it, no trees to protect it. There was nothing.

As the tree branches whipped around, they shredded the tarp that served as my shelter. Sleet and hail sliced through the tattered pieces of what used to be my roof and walls. Every new gust flipped the platform up into the air, threatening to hurl me over the edge.

I was scared. I take that back. I was terrified. As a child, I experienced a tornado. That time I was scared. But that was a walk in the park on a sunny Sunday afternoon compared to this. The awesome power of Mother Nature had reduced me to a groveling half-wit fighting fear with a paper fork.

Rigid with terror, I couldn't imagine how clinging to a tiny wooden platform for dear life could possibly be part of the answer to the prayer I had sent to Creation that day on the Lost Coast. I had asked for guidance on what to do with my life. I had asked for purpose. I had asked to be of service. But I certainly never figured that the revelation I sought would involve taking up residence in a tree that was being torn apart by nature's fury.

Strangely enough, though, that's how it turned out. As I write this at the age of twenty-five, I've been living for more than two years in a two-hundred-foot-tall ancient redwood located on Pacific Lumber property. I have survived storms, harassment, loneliness, and doubt. I have seen the magnificence and the devastation of a forest older than almost any on Earth. I live in a tree called Luna. I am trying to save her life.

Believe me, this is not what I intended to do with my own.

⌒⌁

I SUPPOSE IF I look back (or down, as the case may be), my being here isn't all that accidental. I can see now that the way I was raised and what I was raised to believe probably prepared me for where I am now, high in this tree, with few possessions and plenty of convictions. I couldn't be here without some deep faith that we all are called to do something with our lives—a belief I know comes from directly from my parents,

Dale and Kathy—even if that path leads us in a different direction from others.

Even when I was a child, we hardly lived what people would call a normal life. Many of my early memories are full of religion. My father was an itinerant preacher who traveled the country's heartland preaching from town to town and church to church. My parents, my two brothers, Michael and Daniel, and I called a camping trailer home (excellent preparation for living on a tiny platform), and we went wherever my father preached. My parents really lived what they believed; for them, lives of true joy came from putting Jesus first, others second, and your own concerns last.

Not surprisingly, we were very poor, and my parents taught us how to save money and be thrifty. Growing up this way also taught us to appreciate the simple things in life. We paid our own way as much as possible; I got my first job when I was about five years old, helping my brothers with lawn work. We'd make only a buck or so, but to us that was a lot. I had my share of fun, but I definitely grew up knowing what *responsible* meant. My folks taught me that it was not just taking care of myself but helping others, too. At times, like right now, I have lived hand to mouth. But I knew that sometimes the work of conveying the power of the spirit, the truth as I understood it, was as important as making money. I've always felt that as long as I was able, I was supposed to give all I've got to ensure a healthy and loving legacy for those still to come, and especially for those with no voice. That is what I've done in this tree.

By the time I was in high school in Arkansas, life settled down for us, and I lived the life of an average teenager, working hard and playing hard. I knew how to have fun, and I enjoyed myself and the time I spent with my friends. I was a bit aimless, volunteering for a teen hot line here, modeling a bit there, saving money to move out on my own. I suppose I had the regular dreams of a regular person.

All that changed forever, though, that night in August 1996 when the Honda hatchback I was driving was rear-ended by a Ford Bronco. The impact folded the little car like an accordion, shoving the back end of the car almost into the back of my seat. The force was so great that the stereo burst out of its console and bent the stick shift. Though I was wearing a seat belt, which prevented me from being thrown through the windshield, my head snapped back into the seat, then slammed forward onto the steering wheel, jamming my right eye into my skull. The next morning when I woke up, everything hurt. "I feel like I've been hit by a truck," I said out loud, and then I started to laugh. "Wait a minute, I *was* hit by a truck!"

Although the symptoms didn't surface immediately, it turned out that I had suffered some brain damage. It took almost a year of intensive therapy—much of it alternative—for the information tunnels in my brain to be retrained and rerouted and for my short-term memory and motor skills to return. For a time it was uncertain whether I was ever going to be able to function normally again.

When your life is threatened, nothing is ever the same. I suddenly saw everything in a new light. All the time and space I had taken for granted became precious. I realized that I had always been looking ahead and planning instead of making sure that every moment counted for something. I also saw that had I not come through the way I did, I would have been very disappointed with my empty life. Perhaps because I had injured the left, analytical side of my brain, the right, a more creative side, began to take over, and my perspective shifted. It became clear to me that our value as people is not in our stock portfolios and bank accounts but in the legacies we leave behind.

My parents' legacy began to take hold. I guess I really am the daughter of a preacher through and through after all. Having survived such a horrible accident, I resolved to change my life, and I wanted to follow a more spiritual path. If I was again to become whole—and that meant body, mind, and spirit—I was going to have to find out where I was meant to be and what I needed to do. Lying in bed, still under many doctors' care, I decided that when I was well enough I would go on a journey around the world. I would visit the places that had deep spiritual roots. In those roots, in that common thread of spirituality, I felt, I would find my sense of purpose. The insurance settlement from the accident would help provide the funds. But I couldn't leave the country until the lawsuit I had filed after my accident had been settled. And, of course, I couldn't go anywhere until I got well.

Ten months later, I was finally released from my last doctor's care. I was ready to go. But where? In just two weeks, adventure presented itself. My next-door neighbors, Jori and Jason, whom I had known for years, along with a friend of theirs named Barry, announced that they were setting out on a trip to the West Coast.

"Our fourth backed out, and we really need someone else to help pay for gas," they told me. "You want to go with us?"

I couldn't have been happier. It wasn't the world, but it was a start. We set off in June. Our group was eager to see Washington's Olympic rain forests, but our travel plans were loose. We met a stranger in passing who told us that if we were heading west, we had to stop in Humboldt County and see the Lost Coast and the redwoods, one of the largest undeveloped coastlines left in this country. I was convinced. We changed course and followed his advice.

After stopping at the magnificent shore, we entered Grizzly Creek State Park to see the California redwood giants. When Jason and Jori learned that their dog would not be allowed on the trail, they decided to walk the road that circles the campground and admire the redwoods from there.

That wasn't enough for me.

"There's something about these trees," I said. "I have to get away from all the tourists and the cars and really get out in this forest."

"Well, Julia, we're only gonna stay for a few minutes, maybe fifteen minutes tops," they countered.

"No problem," I answered. "If I'm not back when you're ready to go, just leave my stuff at the ranger's station and tell them I'll be glad to pay however much it costs to store it there. But I'm going out."

As I crossed the highway, I felt something calling to me. Upon entering the forest, I started walking faster and faster, and then, feeling this exhilarating energy, I broke into a run, leaping over logs as I plunged in deeper.

After about a half mile, the beauty of my surroundings started to hit me. I slowed down for a better look. The farther I walked, the larger the ferns grew, until they were so big that three people with outstretched arms couldn't have encircled them. Lichen, moss, and fungus sprouted everywhere. Around each bend in the path, mushrooms of every shape and size imaginable burst forth in vivid hues of the rainbow. The trees, too, became bigger and bigger. At first they seemed like normal trees, but as I leaned my head back as far as I could, I looked far up into the air. I couldn't even see their crowns. Hundreds of feet high, they were taller than fifteen-, eighteen-, even twenty-story buildings. Their trunks were so large that ten individuals holding hands would barely wrap around them. Some of the trees were hollow, scorched away by lightning strikes, yet they still stood. These trees' ancestors witnessed the dinosaur days. Wrapped in the fog and the moisture they need to grow, these ancient giants stood primordial, eternal. My feet sank into rich earth with each step. I knew I was walking on years upon years of compounded history.

As I headed farther into the forest, I could no longer hear the sounds of cars or smell their fumes. I breathed in the pure, wonderful air. It tasted sweet on my tongue. Everywhere I turned, there was life whether I could see, smell, hear, taste, or touch it or not. For the first time, I really felt what it was like to be alive, to feel the connection of all life and its inherent truth—not the truth that is taught to us by so-called scientists or politicians or other human beings, but the truth that exists within Creation.

The energy hit me in a wave. Gripped by the spirit of the forest, I dropped to my knees and began to sob. I sank my fingers into the layer of duff, which smelled so sweet and so rich and so full of layers of life, then lay my face down and breathed it in. Surrounded by these huge, ancient giants, I felt the film covering my senses from the imbalance of our fast-paced, technologically dependent society melt away. I could feel my whole being bursting forth into new life in this majestic cathedral. I sat and cried for a long time. Finally, the tears turned into joy and the joy turned to mirth, and I sat and laughed at the beauty of it all.

Two weeks later, I found out that if I had walked a little farther along the path, I would have been dumped into a clear-cut courtesy of Pacific Lumber/Maxxam Corporation, where these trees, which had taken thousands of years to grow, had been felled in moments with chain saws. Less than 3 percent of these unique wonders were left in the world, the rest turned to housing lumber and patio furniture. Non-profits exist in this country to preserve churches hundreds of years old, but these trees had few established groups to save their thousand-year-old lives from Maxxam's greed. Learning about the clear-cut made me feel like a part of myself was being ripped apart and violated, just as the forests were. These majestic ancient places, which are the holiest of temples, housing more spirituality than any church, were being turned into clear-cuts and mud slides. I had to do something. I didn't know what that something was, but I knew I couldn't turn my back and walk away.

I walked out of the forest a different woman. I certainly felt a calling, but I had some doubts about whether or not the calling was true. On my own now, I decided to return to the Lost Coast to do what I had been taught to do—pray for guidance. I went to a place I'd found on my earlier visit where I had felt an immediate sense of magical power. I hiked down to a special spot nestled between some trees and a stream that flowed out of the King's Mountain Range into the ocean. I sat down and began to pray.

When I pray, I ask for guidance in my life to be the best person I can be, to learn what I need to learn, and to grow from what I learn. Always when I pray, I ask to let go. Letting go is the hardest part.

"Universal Spirit, I wanted to go around the world," I prayed. "I've been wanting to travel ever since I can remember. I finally have the chance, and yet I'm suddenly feeling compelled not to go. Please show me the way."

I believe in prayer, but ultimately the biggest power in prayer for me comes from the willingness to accept the answers. So I added, "If I'm truly meant to come back and fight for these forests out here, please help me know what I'm meant to do, and use me as a vessel." ·

I sat very still for quite some time. After a while, I began to feel completely peaceful about the idea of abandoning my travels in favor of my newly perceived mission. Getting up, I started to walk. That's when I found a crystal—an amethyst crystal. Amethyst is my birthstone. The coincidence was too amazingly synchronistic to overlook. It seemed to me that the spirits approved of my decision.

If I had known what was in store, I'm not sure I would have so readily agreed to follow this urgent call.

⌒～～⌐

I WENT BACK TO Arkansas and settled my lawsuit against the driver of the truck. I sold everything I owned except for my violin, some artwork, and a few photo albums, all of which I packed inside a cedar chest my grandfather had made for me and stored it at my father's house. With the sale money, I bought a backpack, a sleeping bag, and a tent. Then I strapped on the few changes of clothes that were now my total belongings and prepared to return to California to save the redwood forests.

I had no clue what I could do, but I knew I was meant to do something. Even though I didn't realize that I was about to launch into a two-year struggle, a deep and compelling sense told me that I had to walk the path I'd chosen—or rather, the path that seemed to have chosen me. There was a calling, and I would not be at peace until I fulfilled it.

I returned to California in the middle of November 1997. When I arrived in Arcata, a town full of forest activists and students from Humboldt State University, I called EPIC, the Environmental Protection Information Center. On my last trip, this was the group that had told me what was really at stake in the forest. When I asked how to get involved, they gave me a phone number for a base camp.

"This is really the only place you can be of much assistance," they said.

I didn't even know what *base camp* meant. I just knew it had something to do with the forest. So I called from a pay phone outside the town's food co-op, and I let the phone ring and ring and ring. There was no answer. I waited for a while and called back, but there was still no answer. I tried over and over again for two hours. Finally somebody picked up the phone. I gave a brief rundown of who I was and how I'd gotten their number.

"I'm trying to find out where you're located," I said. "I've come to help the forest."

"Actually, base camp is closing," the faceless voice said. "We don't need you."

I felt like I'd gotten punched in the stomach. My eyes started to water. I knew I was supposed to be out here, but I was being told I wasn't needed.

"Isn't there somewhere I can be of help?" I asked, my voice wavering. I was given another number.

"Yes, base camp is closing," confirmed the person who answered my call. "They don't need you there."

"I understand that base camp is closing and they don't need me, but the forests do. There's got to be something I can do. Isn't there some way?"

This newest voice told me about a rally in Eureka, a town seven miles south, and provided directions. When I arrived I saw around three hundred people. The speaker was talking about pepper spray, which law enforcement officers had recently swabbed directly into forest activists' eyes during a peaceful protest. All around me, though, people were laughing and talking among themselves. When the speaking stopped and the attendees started marching toward the courthouse, five people were chanting one thing, three others something else, while ten people farther on sang a song. To me, the lack of unity sucked away much of the march's power.

"If this is the voice for the forests, it's no wonder we're losing them," I thought to myself.

When we reached the courthouse, a speaker addressed the crowd. Then he invited anyone from the public to come up. I kept having this sense that I needed to speak. It almost felt like an invisible person was grabbing my arm, my heart, and my spirit and pulling me toward the front. I kept holding back. I didn't know enough about the issues, and I felt I didn't really belong. But I knew that forests were being destroyed. That was enough. Finally, I walked up to the man with the bull-horn, set down my backpack, and began to speak.

"I'm really glad to see all this joy and happiness in people, but while we're here having fun at this rally there are trees out there dying, falling to the ground, and it's killing me."

I was very emotional and quite uninformed, but in spite of my tears and my lack of knowledge, I continued my impromptu soliloquy. Afterward, a man with dark shaggy hair and a mustache and goatee came up to me. He introduced himself as Shakespeare, the *nom de guerre* he used as an activist. (I would later meet a whole range of activists with fun nicknames, which they used to protect their identities during actions and to bond with one another. Out of respect for their privacy, and because those are often the only names I ever knew, those are the names I will use in this book.)

"You sound like the kind of person who should be involved with us," he said kindly. Clearly he was responding to my passion and not my experience. "Base camp is closing, but it's not closed yet. I don't know any other way for you to get plugged in to the movement. So if you want, I'll help you get there."

We hitched to the much-heralded and elusive base camp, and when we got there I jumped out of the car and promptly sank into six inches of muck. Scattered across a muddy field were tents of all colors, a couple of lean-tos, a large pile of trash and garbage, and in the middle a larger white tent, which seemed to be the central meeting spot. Everything smelled of wet dirt, sweat, and patchouli oil.

Sure enough, base camp was closing down. The major push of direct action in these forests, it turned out, typically kicks off in September and goes through November, the time period when Pacific Lumber is allowed to log the old-growth trees, which are also habitats for threatened and endangered species. At the beginning of winter, however, everybody from base camp returns home or gets involved in other campaigns or heads elsewhere to rest and recuperate. Winter is no time for direct action outdoors. This year was no exception. With the traditional activist season winding down, the people spearheading the movement were no longer around. Those still remaining were definitely not the core activists.

In short, it was chaos. For three days I tried to get connected, but few people seemed to know what was going on. Worse, no one seemed to want me there. Most didn't even care to talk to me. Shakespeare was one of the few who did.

"Don't give up," he said when, discouraged, I finally threatened to leave. "There are tree-sits going on, and I bet if you stick around a little bit longer, we'll get you trained to climb, and then you can go up and do a tree-sit."

A tree-sit, I discovered, is a tactic used in the struggle to protect the forest. Installing human beings around the clock on a platform high in a tree hopefully prevents that tree, as well as those around it, from being cut down while bringing attention to the point where a forest becomes a product. This kind of civil disobedience is one of the few peaceful methods available to the forest movement. So people kept going up trees to try to save them, and I was certainly willing to join their ranks.

Aside from Shakespeare, however, most of the people at base camp avoided me, unless I was cooking a big pot of food, which I did daily. Otherwise, I sat around a lot, looking for some way to become involved. Mostly I waited—alone. On the rare occasion, someone would ask me my name.

"Julia," I'd respond.

"Don't you have a forest name? Everyone has a forest name."

"I can tell," I said with a smile. "But the only thing I can think of is *Butterfly*. But that feels too sissy to me. I've been thinking about something a little tougher, like *Monarch*." But Butterfly had been part of my life since I was about seven, so I stuck with that. Butterfly it was.

After three or four days I was getting tired of sticking around base camp, especially since the chances of my getting trained to do anything were looking slimmer and slimmer. Shakespeare kept urging me to hang on.

"It'll happen. Just wait and see."

Sure enough, one day while we were walking around and talking, a tall, hollowed-out man with gaunt cheeks, deep-set eyes framed by heavy eyebrows, and a chipped tooth showed up.

"I need people to sit in Luna," he cried as he walked the compound. "Can anybody sit in Luna?"

"I will!" I yelled with enthusiasm.

I didn't know anything about sitting in a tree, but I had come to do something for the forest, and, finally, this was something. Almond, as he was known, looked me over skeptically. At the time, he was the person responsible for making sure that the various tree-sits were occupied and stocked with food and supplies. The fact that he'd never seen me before made him leery.

"Do you have any experience?" he asked.

"No, but I'm a quick learner" was my excited response.

"Well, I need somebody to commit for a long period of time," he countered. "I need somebody for at least five days."

I assured him that wouldn't be a problem. Still he hesitated. But there was no one else to accompany Shakespeare and Blue, the only two other people who had volunteered. Grudgingly, he accepted my offer.

And so it began.

My brothers and I in 1980. From left, Michael (eight), Dan (five), and me (six).

With my mother Kathy and brothers Mike and Dan in front of the trailer we lived in for four and a half years, traveling with my father, the evangelist.

Family photo, 1982. From left, Dale, Kathy, Dan, Mike (standing), and me.

TWO

INITIATION

M̲y pack felt heavy. I was not in top shape. As the rain fell steadily, I set out with my companions, Shakespeare and Blue, a somewhat heavyset man in his early twenties. Our guide, Geronimo, a taut, wiry man with long, dark hair, a curved scar on his face, and scars emanating from deep, pain-filled eyes, led the way.

I didn't know much about the forest-activist movement or what we were about to do. I just knew that we were going to sit in this tree and that it had something to do with protecting the forest.

"This is where the adventure begins," Geronimo announced with a laugh when we reached the gate marking Pacific Lumber's property line. I grinned.

"All right!"

We climbed over the gate and started hiking. The first part of the hike was relatively flat, which was great. But before long, the terrain got really steep. So-o-o steep! As one of the tree-sit coordinators, Geronimo had been hiking up the steep hill to the tree everyone called Luna once, sometimes twice, a day. He was in such great condition that his feet barely touched the ground. It seemed like he was floating. I, on the other hand, was sinking.

I wanted to hike fast and not be a burden. So I tried to keep up, but as we rose in elevation, I quickly began to have problems breathing, a typical reaction for me. To make matters worse, my back and shoulders were killing me. But my legs were holding strong, so I tried to focus on them and how strong they felt. Then I tried imagining that I was a bird on a wing so that I wouldn't feel the weight of my pack and the heaviness of my body. That worked for a little bit, and I quit struggling as much.

Geronimo was very patient. He would burst way ahead and then wait for us to catch up. I relished every reprieve, yet each brought a fresh problem. The strenuous, fast-paced, uphill hike had made me sweat. But the air was so cold and wet that every time I stopped to catch my breath, I would turn into an air conditioner and start to freeze. Then I'd take off fast to warm up, run out of strength, stop, and turn into an ice ball again. In life, those things that are really worth achieving rarely come easily. This was certainly no exception.

By the time we'd reached the highest point, the logging road was just one soupy, muddy mess. That's what a solid month of rain will do. The weight of my pack kept pushing me down into the mud until I nearly had to jump up with each step to free myself from the suction. Over and over, the pack would push me down, I would try to spring out, sink back down, stop to catch my breath, and then leap again. The unrelenting effort became almost unbearable.

The trail suddenly grew even steeper.

"My God, I'm not going to make it up this grueling hill. I'm not. I just can't do it. I can't do it," I thought to myself.

The self-doubt seemed to add another hundred pounds to my back, intensifying my exhaustion.

"Oh, Creator, go ahead and kill me now so I need no longer suffer!" I asked, only half in jest.

But the Universal Spirit had something else in mind for me, so I continued to struggle through the muddy mess.

"Hey, stop for a second," Geronimo suddenly exclaimed as we neared the crest. "You see that blinking light up there? That's Luna."

That was my introduction to the Beacon of Hope. Toward the beginning of the tree-sit, activists decided to hang a beacon on Luna so that people in base camp could see it blinking and know that the tree was still standing. The sight of it gave me this great surge of hope, which, of course, is what that beacon was all about. Suddenly, I knew I could make it.

Soon we'd reached the part of the logging road that skirts behind Luna. Only a vague outline of the tree's top could be seen. We howled out to the tree-sitters—a Canadian named Puck and a woman named Zydeco—both of whom had been up in the tree for a few days. They howled back.

"Are you all right?" we hollered. "Do you need anything?"

Hearing that they were fine, we went on to what is called a satellite camp. Satellite camps are usually located in between base camps and tree-sits. They always have a point person who makes sure the tree-sitters are okay and often provides them with cooked food. Bulk foods and supplies are also stored there. That way there's enough room for people and their gear in the tree, since space up there is always limited.

Exhausted by the time I reached satellite camp, I thought it looked like paradise. This wonderful camp was made out of tarps draped over an old, half-burned-out snag (a tree that has died). The camp reminded me of forts I made when I was a little girl.

Unfortunately, the small space was already occupied by two guys who weren't supposed to be there. There was no way that Shakespeare, Blue, and I, along with these two surprises, were all going to fit under the tarps and stay dry. And I was already wet to the bone and freezing cold because I was wearing cotton.

In the activist world of temperate rain forests, the word is that cotton kills because, unlike wool, it doesn't keep you warm once it's wet. I'll vouch for that. I was so cold that my chattering teeth were about to shatter my head. All I wanted was to get my clothes off and climb into my sleeping bag, huddle like a cocoon, and wait for the sun to rise. Which is exactly what I did. But we were all so packed in—with me on the edge—that I spent a wet and largely sleepless night. By the next morning, I was more than ready to check out the tree that I'd heard so much about.

Luna was "discovered" by a handful of activists from Earth First!, a direct-action movement started in 1980 by people tired of the failures of trying to work within the political system. In October 1997, upon learning that an area closely located to the Stafford slide was being logged, an Earth First! reconnaissance group decided to have a look. After checking out the logging operation, then located on a lower slope, the group hiked up farther, since the land above was targeted next. Toward the top of the hill, they saw a gigantic redwood tree marked with blue paint, which meant that it was slated to be cut down.

Deciding that it had to be saved, one young man free-climbed a younger tree growing out of the older tree's main trunk and then crossed over the branches into the main tree. About five days later, a dozen more people hiked back up the hill, laden with ropes and wood scraps from a salvage yard. By the light of the full moon, they built the platform I was about to climb up to. They named the tree *Luna,* which is Spanish for "moon," to commemorate the event, and they began the organic tree-sit.

Tree-sitting is a last resort. When you see someone in a tree trying to protect it, you know that every level of our society has failed. The consumers have failed, the companies have failed, and the government has failed. Friends of the forests have gone to the courts, activists have tried to make consumers aware, but with no results. Corporations have neglected their responsibility as landowners, while the government has refused to enforce its laws. Everything has failed, so people go into the trees.

"I have no other way to stop what's happening" is basically what a tree-sitter is saying. "I have no other way to make people aware of what's at stake. I've followed the rules, but everything I've been told to do is failing. So it's my responsibility to give this one last shot, to put my body where my beliefs are."

Efforts to preserve the coast redwoods date back to the early 1900s, when four society women from Eureka, California, birthed the movement by forming the Save-the-Redwoods League. Dressed in fur-trimmed coats and fancy hats, they wrote letters to politicians as well as to a number of renowned and moneyed naturalists. Then they took their cause on the road, just as today's protesters have taken their cause to the trees, in order to promote public awareness. Without these past and present activists, more of California's ancient forests would have fallen long ago.

Though Luna started as an Earth First! tree-sit, I wasn't—and still am not—part of that group. In truth, I didn't even know what Earth First! was until my second stint in the tree. All I wanted to do was have a purpose and direction in my life. Helping save the ancient forest seemed like the urgent answer to a question I didn't even know I was asking. I really didn't care what group sponsored which actions.

From the very start, sitting in Luna gave me a sense of purpose. Here was something I could do to make a difference. Of course, I had no idea that this would mean abandoning the ground for two years. And I had no idea what the forces of nature—and a company called Pacific Lumber/Maxxam— would soon unleash upon my head.

Had Pacific Lumber remained the family-owned and -operated mill it was for over one hundred years, there probably would have been no need for a Luna tree-sit. From 1885 to 1985, everyone who worked for the mill was either family, friends, or neighbors. The Murphy family who owned it also owned Scotia, the town directly north of Stafford where the mill is located, and the company had operated under a policy of long-term sustainable forestry. They knew that if they cut too much, they would be cutting their grandchildren out of a job and a future. So they cut at a modest rate, taking large trees but leaving a diversity of plants and trees, which allowed for those trees that were left to thrive in the sunlight and the forest ecosystem.

By the fall of 1985, mostly because of their sustainable logging practices, Pacific Lumber still had an incredible amount of assets in standing timber. Enter Charles Hurwitz, a Texan who heads Maxxam Corporation and is notorious according to press accounts for taking over businesses with undervalued stock, then liquidating their assets. With the help of Drexel Burnham Lambert, Michael Milken, and Ivan Boesky, who were later sent to jail for insider trading, he financed high-interest junk bonds, which he used in a leveraged buyout.

Hurwitz incurred an $800 million debt in his takeover of Pacific Lumber, which up to that point had been virtually debt free. To pay that debt, he sold off many of the company's assets, and aggressively stepped up the rate of cutting, thereby destroying the sustainability of the forest, harming the environment, and threatening many of the species—including humans—who lived there. He also used a $64 million dollar pension fund to pay off junk bond debts and faced charges from pensioners alleging he had taken their money.

Pacific Lumber's press releases claim that they are taking good forestry into the twenty-first century. But it is documented thoroughly that the practice of clear-cutting takes away nature's ability to absorb water and lock in precious soil. So the water saturates the ground, which no longer has all those wonderful root systems and trees to suck it up like sponges and to hold the ground structure together. Steep hillsides then wash away in walls of mud that destroy streams— and species like the coho salmon, which is threatened with extinction because the silt-congested streams choke them and their eggs—or in slides that wipe out hillsides and towns like Stafford.

To Hurwitz and his bondholders, though, trees and land are simply potential dollars, and anything that threatens those dollars, like laws or environmental activists, are seen simply as hindrances.

~~~~~

OF COURSE, none of that entered my mind when we saw Luna the next morning. I was working purely on instinct. Not until months later would I educate myself on every matter relating to sustainable forestry. But for the moment, all I could

Looking up at Luna.

focus on was the tree, and her two hundred feet of height, which I was about to scale.

Climbing this tree looked impossible. At night, I hadn't seen how enormous she was. But when the sun came up and I saw where the Beacon of Hope had come from, I was overwhelmed by the enormity and dignity of this tree. Luna is not a typical redwood tree, which grows straight like an arrow. Those are impressive enough. But Luna sits at the top of a cliff near the crest of a steep mountain and can be seen for miles. She gets buffeted by high winds and has stood through forest fires. She has been struck by lightning, so black scars are burned all the way up the trunk. At the bottom are two caves lined with charcoal, one on the uphill side and one on the downhill side. The caves extend almost completely through the trunk, but Luna still holds strong. Redwoods are amazing. They can be hollowed out by lightning fires, but their thick, moist bark preserves them. I would be taking much instruction in the coming months from Luna—a thick skin for survival just one among many.

That morning as I gazed up at the redwood's great trunk, I noticed a rope about the width of a dime snaking down its side. It seemed to come from somewhere way up in the sky. But no matter how far I bent back, I couldn't see the point on Luna from which it emanated.

"I don't want to climb that rope!" I thought to myself, having quickly compared the width of the tiny rope to Luna's trunk, which measured over twelve feet in diameter. "There's no way that little rope is going to hold me up!"

Despite my reservations, Shakespeare's tree-climbing lesson followed almost immediately. It would do little to calm my fears. First he showed me how to put on this harness, which had *duct tape* on it! From the beginning, Luna has been the riffraff tree-sit, especially since many of Earth First!'s people did not want the Luna tree-sit to happen. At the time, all the funding and all the resources were going out to activists involved in setting up the Liberty tree-sit, their baby of the moment, which had cost thousands of dollars to build and thousands of dollars to maintain. All the best food, all the best gear, and all the best activists went to Liberty. Luna got the scraps, including the harness held together with duct tape, which I had just put on.

Actually, I was so focused on what Shakespeare was telling me about how to climb this tree that I barely noticed the jerry-built harness. Knowing that my life could depend on what I was about to learn, I gave it my full attention. My lesson, however, took only about three minutes.

"This is a prussic knot," Shakespeare told me, showing me how to wrap an even skinnier cord about the width of a pencil three times around the climbing rope.

"This is how you move it," he said once I demonstrated that I'd mastered the knot-tying technique.

He handed me the harness, which is a belt that goes around your waist. Loops are attached to it that fit around your legs like belts. It also has metal clamping hooks called carabiners.

Then he explained that by looping the prussic line around the main climbing line, it grips the climbing rope like a fist when pulled down by the weight of your body. When you take your weight off, you can loosen the knot and move it. One end of the prussic rope grips the climbing line knot, the other loops around one of the metal carabiners at your waist, where it gets locked down.

Attached to another carabiner is a rope with a loop on the bottom for your foot and an extra prussic knot that also attaches to a locking carabiner on your waist. So it basically looks like an upside down *T,* with one end attached to the climbing rope, the bottom looped around your foot, and a shorter rope crossing those that are attached to your waist.

To climb, you put your foot in the loop and stand straight. That pulls the slack out of the rope and also locks that prussic down, giving you room to scoot the one called the chest prussic as far as your arm will reach. You lean back on that one, which locks the grip knot on the climbing line, then loosen the one on your foot by wiggling your hand. You push that one up as far as your leg can go in the air. Then you stand on it, and the weight of your standing body tightens the knot. So you go back to the chest prussic, loosen the knot by wiggling it, slide it as high as you can, lean back on it, and repeat the process all over again.

That was the extent of Climbing 101. Scared, my legs shaking, I began to inch my way up the tree. At first all I could think about was moving my knots correctly and getting into the rhythm: *loosen, move up, stand, loosen, move up, lean back, loosen, move up, stand, loosen, move up, lean back.* I was developing a whole new appreciation for earthworms.

When I got somewhere between fifty and seventy-five feet off the ground, however, the fragility of my rope and harness suddenly got to me. I was now high enough off the ground to hurt myself really bad if something went wrong. The wind had picked up, and I was swaying a lot.

I felt my breath catch in my middle.

"What am I doing?" I thought with a rising sense of alarm. "This rope can't hold me!"

That's when Luna called out to me, causing my attention to veer from the little thin rope to the tree's massive trunk. And instead of feeling afraid, I simply started examining this incredible tree I was climbing.

❧

THE BRANCHES ON the lower part of Luna have all broken off. The trunk sports gnarled burls shaped like different animals: one looks like a hippopotamus, another looks like a monkey, lots of them look like a human female's breasts. Its bark is soft and shaggy with amazing, swirling designs. The patterns keep changing, like when you were a kid and you lay on your back in a field and picked out different shapes, forms, animals, and faces in the clouds. Luna looks just like that.

At about eighty feet up, the tree splits in two. After three or four feet, it grows back together for a few feet, then it splits again into two massive pillars. Twenty feet higher, some incredible limbs grow out from the trunks. Each one is huge, the size of a twenty-year-old tree. These massive branches intertwine

© THOMAS DE SOTO

*I may be smiling as I look at this rope, but I sure
wasn't laughing the first time I saw it.*

into one gigantic mesh, one growing into the other, until you
can't figure out which branch is growing out of which trunk. It
reminds me of the jungle gyms I played on as a little girl.

Despite my efforts at distraction, I began to experience a
sense of panic that I couldn't shake. Though I hadn't paid
much attention to my harness on the ground, its compromised
condition now seemed like a vital detail.

"You're wearing a harness with duct tape!" screamed my brain. "You're using little ropes the width of a pencil, and your main lifeline is about as big around as a dime. What in the world are you doing?"

The only answer seemed to lie in the very tree I was trying to ascend. I closed my eyes, put my forehead, my hands, and my feet against Luna. I imagined my energy flowing down the trunk, through her roots, into the ground. That helped me feel grounded. I felt the massiveness of her trunk and this deep, connected energy.

"I'm going to be okay," I thought. "I just need to climb. Luna will take care of the rest."

After that, I was just fine. I kept going. When I needed to catch my breath, I'd find a branch big enough to sit on. I definitely wasn't comfortable enough to just hang from the harness.

At about a hundred and fifty feet, one of Luna's two pillars ends abruptly, but the other pillar continues to a height of two hundred feet. Not far from the top there's a cave burned out by lightning, and sprouting from the thin sliver of trunk surrounding the cave is a lot of new growth. Suspended from this new growth, one hundred and eighty feet in the air, is the platform.

The main climbing rope, however, ends well before that, a fact that no one had bothered to mention. So I was at the top of the line, but the sheltered platform still lay another twenty to thirty feet above me. It seemed like a good time to take a rest and get my bearings.

A view of clear-cutting.

As I looked out, I saw everything we are fighting for and everything we are fighting against in one glance. Far off in the distance lay the Headwaters forests, which were still pristine, and beyond them snow-capped peaks. Closer in, however, lay destruction and horror. I saw a patchwork of burned-out carcasses of hillsides, ripped bare, and, lining the highway, huge stacks of redwood timber—a forest turned horizontal. Directly left of me gaped a wide yawn of mud carved out of the hillside as if by a serrated ice cream scoop. If I'd crawled out to the tip of Luna's outer branches and jumped, I would have landed in the upper reaches of the slide that had decimated the town of Stafford the year before. In the distance, I noticed smoke pouring out of the chimneys of the Pacific Lumber Company, the mill that takes these beautiful forests and spits them out as lumber.

"Oh, my God, what I could do with a good rocket launcher from here," I blurted without even thinking.

"I heard that!" pronounced a voice from the platform above. "That violates the nonviolence code!"

I started laughing. I would never do something like that, but it was my instinctual response. I didn't even know what the heck the nonviolence code was! As far as tree-sitting and trainings and codes went, I knew nothing.

I would learn a lot, though, in the days, weeks, and months to come.

THREE

GETTING TO KNOW YOU

I still had to get from the end of the rope to the platform some twenty-five feet higher up the tree. Scampering around the tree's branches like a little monkey, Puck, the tree-sitter we had come to relieve, told me what to do next.

"Those are called lobster claws," he yelled, pointing out a rope in the shape of a V, its two loose ends punctuated with carabiners.

I hooked the point of the V to the harness around my waist and picked up the ends of the rope, placing a lobster claw in each hand. The idea was that, with each move, I would clamp those pincers onto the rope that wound around the branches leading upward.

"This way you can free-climb," Puck explained. "If you slip and fall, you'll only drop about a foot instead of a hundred and eighty feet. That might scare you as bad, but at least it won't kill you."

With that small comfort, I lobster-clawed my way to the platform branch by branch. By the time I reached the platform opening, I was so beat that my muscles were shaking. It had taken me only fifteen to twenty minutes to make it from the ground to the top.

"You *flew!*" they exclaimed.

"Really?"

"You made it in fifteen minutes!" Zydeco announced in wonder.

"It felt like an hour," I replied. Then I corrected myself. "Actually, it felt like a lifetime!"

While we waited for Blue and Shakespeare to climb up, Puck and Zydeco showed me around. It didn't take long, since the platform only measured six by eight feet. A second platform, measuring four by eight feet, was suspended on the other side of a major branch, which also supported a hammock. That smaller platform held food, water, supplies, ropes, and tarps. The hammock later became Luna's library, from which I would thoroughly educate myself about the forest and its preservation in the months to come.

"We really want to stay up one more day," Puck and Zydeco confessed after I'd looked around.

"Well, *we* came all the way up this hill to relieve you guys, and now it's time for you to go down so that we can come up," I thought.

But they really wanted to stay another day, and they were in the tree. It was their call.

Frustrated, I prepared to rappel down and spend another cold night in the satellite camp. First, of course, I had to learn how to rappel.

Puck explained, "Look, if your hand's behind you, you're not going to go anywhere. If you move it out to the side, the roping will slide through. You can decide how fast or slow you want to go. There's so many branches that you won't lose control."

That was it. Not much more training than I had gotten on the way up the tree. He basically said, "Trust me and do it." From the start, when I sold everything in Arkansas to come out to the redwood forest, this whole experience had been nothing but one step of faith after another.

"You'll be fine," he assured me.

And once I started down, I was.

BY THE NEXT morning when Puck and Zydeco were finally ready to come down, the worst rainstorm I'd seen thus far set in. Because Blue had such trouble climbing up the tree the day before, we decided that he should stay down. Shakespeare, who had climbed into tree-sits but never stayed overnight, went first, followed by Major Resin, a skinny guy with unruly, curly, blond hair, a straggly mustache and goatee, and sparkly, giggly eyes, who had unexpectedly shown up the night before.

By the time I was finally able to start climbing, the wind was blowing so hard that for every foot up, I got blown three feet to the side.

"Tell me again why I'm doing this," I muttered as I struggled to climb. "What's the point of all this?"

Because I'd had such success with my first climb, I'd already conquered my fear of making my way up the equivalent of an eighteen-story building on a rope the size of my forefinger. The wind battering me and the tiny rope, however, added a whole new element—both in difficulty and fear. Once again, though, whenever I started feeling panicky I'd close my eyes, put my forehead, my hands, and my feet up against Luna, and feel the solidity of the tree.

When I reached the top of the line, I had to wait for Zydeco to descend before continuing on to the platform. That was the only way I'd be able to pull the rope up. (A rope left hanging could be climbed or cut by people invested in seeing the tree-sit end.) The wind howled around me, blowing the branches wildly. I was soaked to the bone, and my teeth wouldn't stop chattering. All I could think about was how freezing cold I was. My hands had started to go numb, and with a sense of rising dread, I imagined getting hurt because I was just too cold to hold on anymore. My fear made me shake even harder.

All Zydeco could think about was how freaked out she was because the wind was so bad. At the top of the tree where she was, the smaller branches whip around in the wind. Her eyes were big as saucers, and she kept apologizing.

"I'm sorry, I know it's taking me a long time, I'm really, really sorry. I'm sorry!"

With each wind gust, she would grip a branch so tightly you

could see the whites of her knuckles. Her fear took my thoughts off myself and my own fears of slipping and falling.

"It's okay, sister, take your time. Don't rush, I'm going to be fine. I want you to be safe. That's what I'm concerned about the most. You take all the time you need, you just be safe. You're going to be fine, we're going to get you down this line. You're going to be on solid ground, and you're going to be A-OK."

After what seemed forever, Zydeco finally hollered to signal that she had safely touched down and that the line was cleared. I pulled the rope up, which seemed to take another forever, tied it off, and finally made it to the platform.

By that point, the platform seemed like the most wonderful place on earth. As I kneeled at the entrance taking it in, a huge gust of wind lifted up one whole side of the tarp like a sail, flinging me toward the front of the structure. I managed to grab onto a branch, which is the only thing that kept me from flying right over the edge.

"Whoa!" I yelled once I found my voice again. "I'm going to do something about that!"

The wind had ripped the top part of the connection to the top part of the tarp. So Major Resin, the most experienced among us, climbed outside into the branches to fix it.

"He's insane!" I thought, not realizing that I would be doing much the same in the not-too-distant future.

Together we fixed the tarp. While Major Resin worked from the outside, I took pieces of twine and rope from a little bag of scraps that Zydeco and Puck had shown me the day before and wove a web on the inside of the tarp so it wouldn't act so much like a sail. It needed to give, so it wouldn't break the branches, but it certainly couldn't continue to billow the way it was.

My rigging worked pretty well. Once our task had been completed, we tried to situate ourselves in our small space, taking out the gear we needed and then stowing the rest of our stuff out of the way. Then I fixed dinner.

By the time we'd eaten, all I wanted to do was figure out who was going to sleep where so that I could get into my sleeping bag and relax. Because Major Resin had done the most amount of scary work, he was given the right to pick the spot he wanted. He chose the hammock, which was fine by me since I can't sleep in a hammock because of my bad back.

Shakespeare and I shared the platform, with me on the outer edge, since Shakespeare felt more comfortable being on the inside, by the hammock and the big branch that comes through the middle of the flooring. No one was tied down, and just rolling a bit could have pretty much ended the story. I settled down and tried to sleep. But the wind was making a lot of noise in the tarps, *flap flap flap flap flap,* and the platform was swaying back and forth and back and forth.

For most of the evening, I'd managed to focus on what needed to be done—getting Zydeco down safely, fixing the tarp, making the meal, and settling in. Suddenly, all I had to focus on was the energy of the storm.

Though the wind finally slowed, slumber didn't come easily. I don't sleep well next to others, and we had three people piled into a tiny space, two of us scrunched side by side with less than three feet of turning space apiece. Finally I reached the point of sheer exhaustion when my eyes wouldn't have stayed open any longer even if I'd wanted them to, and I finally drifted off to sleep.

Over the next days, the weather didn't improve much, so apart from a little bit of climbing around, I spent most of my time "inside." I read the couple of books that were in the tree, drew, and did a lot of tidying and organizing to make the platform like a home. Since I love to cook—it's a soothing therapy for me and a way of nurturing and comforting others—I also cooked us every meal on the little propane burner we had and then did all the dishes and cleaned everything up. That took a good couple of hours each time. Everything had to be thought about differently. We had hiked all our food and water up the hill. So both commodities were scarce. I did the dishes by boiling about an inch of water in a pot with some herbal soap. I then rotated and spun all the dishes around to make sure they were sterilized. Then I scrubbed them with a plastic scrubby to get any food off. Dumping the water overboard, I boiled another inch of new water for rinsing. Because we had not yet come up with a water collection system, water had to be treated as preciously as possible. My days were strangely packed.

Had I been by myself, I probably would have explored more. But given the company, our stay turned into a hanging-out-with-the-guys experience. We spent a lot of time talking and cracking jokes. Instead of getting to know Luna, we got to know one another.

And believe me, living in such intimate quarters we got to know each other really well. As a woman, I faced some added challenges, like having to go to the bathroom with two men *right* there. That was hard for me because I come from a strict, Puritan kind of background, and peeing in a jar three feet away from two guys was a bit intense! So was trying to stay focused and not let living all together in this small space bother me.

ON THE SIXTH day, as an activist who had come to say hello prepared to head back down the hill to base camp, I was more than ready to accompany him. I knew that Major Resin and Shakespeare would be fine by themselves in the tree. Besides, the cold and wet had caused some nasty flare-ups in my right shoulder and neck, neither of which had completely healed following my accident. So I decided to go down a day early, hoping that a little heat running back through my body would help work the pain out.

At the base of the hill, I got a ride into town from a guy I'd never met. To him, I was just another of those scraggly faces, covered in mud, sopping wet. He took me to the jail support house, where activists just released from incarceration can go to regroup. There, I got to take a shower and clean my clothes and sleep under a roof.

That was all I needed.

"You don't happen to want to go back up, do you?" Almond asked the very next day.

He wanted someone to replace Shakespeare and Resin, who were coming down. I felt completely rejuvenated: my clothes were clean, my body was clean. I was ready to go. And up I went.

This time, I bonded with Luna. The weather had improved some, so I climbed around on the tree when I could and started getting used to maneuvering around.

The rest of the time I talked with the only other person in the tree. Mike, a short man with reddish blond hair and a beard that he played with all the time, told me about attempts made to physically force him out of a tree-sit. His stories about being pepper-sprayed, where law enforcement agents swabbed the irritant directly into activists' eyes in an attempt to get them to abandon their locked-down positions, filled me with horror and sadness. Then, in his quiet, monotone voice, he told me that he was participating in the Earth First! lawsuit.

"What's Earth First!?" I asked.

"It's the group you're a part of!" he answered.

"I'm not a part of any group!" I retorted, somewhat indignantly.

"Well, they're the ones who started this tree-sit."

That was the first I'd heard of them.

"So you haven't had any nonviolence or backwoods training?" he asked. "That's against the regulations."

I started laughing.

"You have to have rules and regulations to sit in a tree? I just came because there was no one else."

I HAD TOLD Almond I would stay however long he needed me to. On the fifth day, however, I got sick. Really sick.

I started feeling lightheaded and queasy. I tried to blow it off, but I got sicker as the day went on. Extreme hot and cold flashes were followed by extreme temperature spikes. My stomach felt like it was turning around and around, getting ripped out of me, tied in knots, and shoved back in. Then I got the shakes, the point where I started going into convulsions. I ricocheted back and forth, waves of shivers wracking my body.

Poor Mike freaked out. He was so afraid.

"Has this ever happened to you before?" he asked frantically. "What do you want me to do?"

There was no one at satellite camp by this point; Geronimo had gone back down the hill. I felt worse for Mike than I did for me. I knew he was panicking, because I would have been, too, had the roles been reversed. As it was, I was too sick to panic on my own behalf.

"Man, don't worry, I'll be all right," I said. "We'll just stay here until someone comes to relieve us."

By the next day, Thanksgiving Day, as it turned out, my convulsions had stopped. Though the hot and cold flashes had returned, I was no longer tossing about shivering.

We didn't have any communications at all—no cell phone, no walkie-talkie, nothing. If Pacific Lumber had come, if anything else had happened, we would have had no way to get hold of anyone except by physically going down the hill.

Almond showed up with a batch of new tree-sitters that afternoon.

"We've come to see if you guys wanted to be relieved," Almond yelled up.

"Oh, my God, you don't understand how glad I am that you're here," I yelled back.

When I came down, he looked at me. "You don't look so good."

"I'm so sick, Almond!" Then I whirled away. "Wait a minute, don't look at me."

I took off, ran behind a tree, and vomited. When the world stopped spinning, I came back.

"Are you all right?" he asked.

"Let's just get out of here," I answered.

We started hiking down the hill. I almost passed out four times. My vision would start to fade, and I'd feel top-heavy like I was going to topple over and fall head over heels down the hill. But we never stopped once. I just wanted to be down off Pacific Lumber property, where I could be safe and be sick. So I just kept going.

It took me three days to start feeling better and another two and a half weeks to recover completely. It turned out that I was under attack by two different viruses, one of which had started moving in on my kidneys.

During that period, Almond looked after me and helped

me get to the doctors. We spent our time together talking about strategy and activism, about what we wanted to do, and about where we felt this effort with Luna should go. I picked his brain, because though I had sat in Luna twice at that point, rotating in and out of the tree for a week at a time was not enough for me. I felt like I needed to be doing more. I just didn't know what.

During my recovery period, I got involved with tabling, which meant setting up a table with information about forest preservation efforts and soliciting donations. The tabling materials, which consisted of some pictures and pamphlets, were falling apart. So I spent about two and a half weeks helping to design and lay out new posters, whose pictures and captions provided people with a self-guided tour of forest issues. Along the way, I learned a lot more about what those issues were. I also learned that in almost every tree-sit ever attempted the tree had been lost.

After watching Almond struggle day in and day out to line up tree-sit volunteers, who often could sit for only a couple days at a time, I could see why. Too much energy went into getting people up and down the hill—and up and down the tree. It didn't make any sense.

Luna was about to come under attack once again by Pacific Lumber as they moved into the upper regions of their timber harvest plan. And there were few folks left to sit in her and save her.

The answer seemed obvious. At least to me.

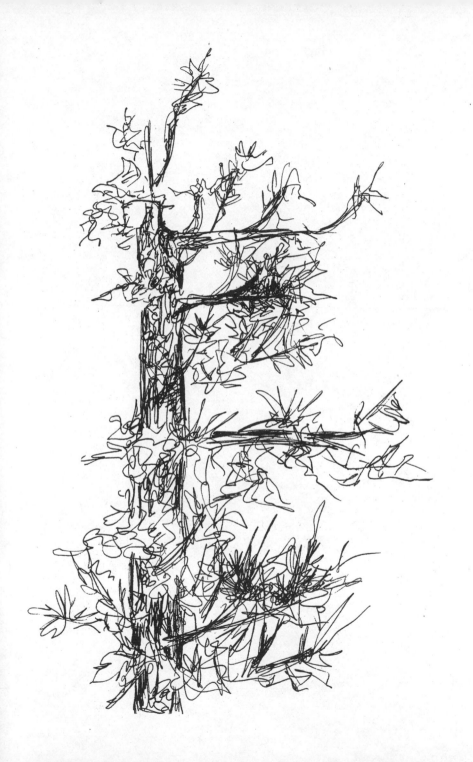

FOUR

CALL TO ACTION

It was December, and I was still recovering from my violent illness, but the cutting had started on the upper part of the timber harvest plan. Everything was in shambles. The people who were up in Luna said they were coming down, yet no one had agreed to take their places. It seemed that no one knew who was up, who was down, why they were up, when they needed to come down, or who would go up next.

We knew we had to keep someone in Luna at all times or she would be cut. The blue mark on her trunk told us that. And the cutting was getting closer to her daily. With most activists either worn out or headed home, however, the pool of candidates was slim. The winter exodus also meant that no one remained to monitor communications from the tree, and there was no phone. All this chaos almost cost us Luna right there.

❧

AT THE BEGINNING of December, just after I came down with kidney problems, three relatively new volunteers had offered themselves for the tree-sit. One of them, a cherubic young woman in her twenties, panicked when Pacific Lumber loggers started an intimidation campaign. Lacking a way to communicate with anybody other than the two people in the tree with her, she came down almost immediately. I understood her fear. The loggers were threatening to blow up the tree, and she was two and a half miles away from anyone who could help. Terrified for her life, she just couldn't take it anymore.

The two remaining tree-sitters—Nature Boy, a sixteen-year-old working on a school project, and a Finnish logger named Seppo, who practiced sustainable forestry in his homeland (which his country mandates and strictly regulates)—both needed to come down as well.

We had to move fast.

"Look, Almond, why don't I go up for three weeks to a month, because at least for that length of time you don't have to worry about whether there's anyone in the tree," I argued to my new friend. "I can pack up enough stuff to keep me alive for that long. It won't be that difficult."

⌒⌒⌒

I HAD ALREADY developed a great affinity for Luna. I had to get back up there. I certainly couldn't let her fall without doing my best to protect her. Almond liked the idea so much, he decided to come along.

Over the next few days, we ran around town buying bulk foods like couscous, oatmeal, farina, dried fruit, instant soup mixes, fresh vegetables, and spices. (If I am the one cooking, I have to have seasoning. I can't stand bland food.) We were given an old video camera and, after a huge struggle with Earth First!'s powers-that-be, managed to wrangle a cell phone. In addition, of course, we packed our clothes and our sleeping bags, along with batteries for all our electronic equipment and tapes for the camera. We also had to carry up our water in those days. So by the time we strapped on our backpacks, each of us was hauling a good hundred pounds up the mountain.

We finally got under way on the evening of December 10, 1997. At that time you still had to hike up at night; by day, you risked getting arrested for trespassing on Pacific Lumber property. We hadn't been on the logging road long when we saw people coming. Instantly, we ditched off the road to get out of sight.

Trying to move quickly yet quietly, I tripped over a vine and crashed face first into a patch of swampy water. We had barely gotten started, and I was already soaked. I was also stuck in the ditch because my heavy pack had slid over my head. I couldn't get into the right position to stand up. It took quite a bit of maneuvering before I was finally freed. Now I know how a turtle must feel.

We waited in tense silence for what seemed like an eternity. My ears and every nerve in my body strained for sounds. Finally Almond felt it was safe.

"No! No!" I hissed as he started to walk out. My fear-sharpened senses had detected a stray noise that didn't belong.

The logger coming down the road must have heard me whisper to Almond. He stopped. Then he forged a little way into the bushes directly across from us. Luckily it was dark, so he couldn't see much of anything. Still, the moon was bright; I worried that he would spot my bright blue backpack. Thankfully, he walked on.

Savoring our escape, we were just about to set out again when I heard somebody else.

"Wait!"

Then I recognized Seppo by the hat he was wearing. "Seppo! Seppo!" I whispered loudly, jumping out to greet him and promptly scaring the living daylights out of him. After a quick chat, we said our good-byes, and Almond and I resumed our hike.

We were so loaded down that it took us three and a half hours to make it to the base of the tree, usually a two-hour trip.

As we approached, Almond and I heard two voices from on high. We were supposed to be relieving the last remaining tree-sitter, Nature Boy. Finding another person up there was really frustrating, since Almond and I could have used the extra day to prepare—and an extra dry, warm night—had we known that we weren't immediately needed.

The tree's current residents lowered down the climbing line and the rope that would hoist our packs up. Attached to that rope was a harness. Within minutes, Almond started climbing.

There was supposed to be an additional harness for me, but it turned out that Seppo had taken his with him. Jump Shot, a hyperactive, caring man in his own little world, and the owner of the unanticipated voice we'd heard in the tree, had been scheduled to go back down with Seppo. Instead, he had decided to climb up, which meant that he had the harness that would have been mine.

While waiting for Almond to send the harness back down for me, I helped get our gear into the tree. Between the wind blowing and the inky darkness, however, the stuff kept getting jammed between the branches. Each time the guys up top would lower the gear back down, and I would have to reposition it on the rope before trying it all over again.

An hour and a half after reaching the base of the tree, we got the last of the provisions up. By then it was midnight. Finally, I was able to put on the harness and ascend Luna. It seemed an exhausting eternity before I finally reached the top. When I got there, I untangled myself from the harness and looked around for a place to collapse. I expected Jump Shot to go down so there would be enough room to sleep. Before long, however, I realized that he had no intention of doing so.

"Oh, God," I thought.

Almond was in the hammock at this point, crashed out from the hike and climb. The other two had settled in as well. All I wanted to do was stretch out in my sleeping bag and rest. But Jump Shot measures six-feet-five, and Nature Boy is also tall and lanky. Then there was me, five-ten, so that meant three really tall people crammed into a space that measures about four by seven feet.

There was no way I was getting any sleep squished between Jump Shot and Nature Boy. Number one, if I can hear people breathing, it's very hard for me to sleep. Number two, I shattered my left arm and elbow in high school, so I have to sleep with that arm crooked. Otherwise it ends up hurting really bad, or the elbow slides out of its socket. But I didn't even have enough space to lay both shoulders down, let alone crook my elbow.

I tried to lie on my side in order to fit between them, but I couldn't sleep. So instead, I scooted to the very edge of the platform down by their feet. Since their feet took up less room than their heads did, I had just enough room to lie on my stomach with my elbows bent. I lay at a right angle with my legs dangling across two-by-fours that connect the platform's entryway (the original platform builders ran out of scrap wood, so there's an opening), while my feet hung off the platform completely.

The next morning, after a nearly sleepless night, my whole body was in knots. In high spirits, Nature Boy and Jump Shot laughed and cracked jokes. As the morning wore on, I got increasingly nervous.

"You guys really should go, because the loggers are going to be here any minute," I warned.

They continued their banter instead. Finally they readied themselves to rappel down. That's when I noticed the Pacific Lumber truck parked on the logging road landing that runs by the head of the slide. Instinctually, I knew something was wrong.

"PL truck!" I yelled.

Right then, *whack! whack! whack! thump! thump! thump!* sounded against the tree. I could feel the vibrations coming through Luna. Loggers had started cutting on her with an ax. Unsure about what was happening, I assumed the worst.

"They're going to blow the tree up!" I thought.

Since the loggers had threatened to do just that, the notion wasn't far-fetched. Nature Boy screamed in alarm then started cussing. Jump Shot chimed in.

"Do you believe in Jesus? If you believe in Jesus, then you'd better stop destroying God's Creation or you're going to hell, 'cause cutting down trees is a sin," he yelled. "Jesus loves you, but he doesn't love what you're doing. So you'd better stop!"

If we could have played Earth First! organizer Darryl Cherney's song "You Can't Clear-Cut Your Way to Heaven," the scene would have been complete.

By then, Nature Boy and Jump Shot were about thirty feet lower than I was, having climbed down to where the rope starts. When the chaos erupted, they climbed even lower. Almond and I had remained up top. Since I didn't have one of the harnesses, I wasn't going anywhere.

In those days, I didn't stray far from the platform without being securely attached to a rope—something about being a hundred eighty feet up in the air. On the platform, however, we couldn't tell what was going on below. All we knew was that the loggers were hitting Luna and that they had cut off the baby trees that grew from her trunk.

"I'm going out there to see what's up," Almond said.

Though he had never sat in a tree before, he did have a harness on. I offered to go in his place, on the condition that he give that up.

"No, I'm going," he reasserted, grabbing the video camera and starting to climb.

That's when he spotted Climber Dan in the other tree.

"Oh, my God!" I cried out at the news.

Though I was new to the movement, I'd already heard many stories—none of them good—about Pacific Lumber's Climber Dan, a mythological figure in the forest struggle. A Pacific Lumber climber who originally worked setting lines to trees when they needed to be felled in certain directions, or felling them himself, he became an expert in taking people down out of tree-sits once activists came into the picture. Word had it that he had gotten tired of dealing with the whole scene and became aggressive. Aggressive with someone's life up in the air is not okay, especially when it's your life he's been aggressive with. So the idea of his being anywhere near us filled me with terror.

True to his name, Climber Dan scaled the tree quickly in his spiked shoes, cutting branches as he went in order to be able to girth the tree with his ropes. By then, my imagination was keeping pace with him, and careening wildly all over the place.

"These guys coming to get us," I said fearfully to myself. "They'll arrest us or hurt us, one of the two, maybe both."

Climber Dan was headed for a line that connected Luna to

a nearby tree. Nature Boy started toward it as well, intent on defending the fort, but Almond got there first.

"I'm going out on the line!" he yelled.

Traverse lines are scary. When you step on them, they bow down—way down—and sway. Suddenly, you're suspended high above the forest floor on two tiny ropes that are waving back and forth and bending with each step. Almond had never been on a traverse line. But he climbed out anyway with the video camera we'd hiked up, wearing a minimum of safety gear. Midway across the line, Climber Dan confronted him.

"You guys are insane!" he screamed at us.

"No, you're insane," countered Almond, starting to roll film in what turned out to be a defective camera. "You're the one cutting down the forests. You're right next to the mud slide. Can't you see the consequences of your ignorance and destruction?"

His words didn't even register.

"I'm going to get you," threatened Climber Dan, coming to the traverse lines.

Almond quickly pointed out that the line hadn't been designed to hold two people.

"You'll kill us both if you come out here," he said.

Climber Dan roughly assured Almond that he had another plan. Upon realizing that he couldn't cross over to get us, he had decided to cut the line. This way, the loggers could cut the other tree down and be able to control its fall.

"You're not going to cut the line, are you?" Almond asked,

as Climber Dan prepared to slice the rope upon which Almond still balanced precariously.

"You better believe I'm cutting the line."

"No! Please don't!" Almond pleaded. "I'm not experienced in all this. I don't have my safety lines on."

"I know you know what you're doing. I know you Earth First!ers. You're professionals," retorted an increasingly irate Climber Dan. "Don't try to lie to me. I'm not going to put up with your bullshit anymore."

"No! You don't understand," yelled Almond. "You're going to kill me!"

Climber Dan refused to listen. "Don't try to mess with my head," he threatened. "I've got it all figured."

Almond began to backtrack toward the tree. He still had three feet to go when Climber Dan cut the traverse line that had connected the two trees. Devoid of safety lines, Almond dropped—two and a half feet. Because he had retreated, a solid branch prevented him from plummeting over one hundred feet to his death.

By this point, I was in a complete panic. With all the yelling and screaming going on, and Almond's close call, the one thing I was all too aware of was the fact that I didn't have a harness. I wasn't comfortable with free-climbing at that point, so I could go only about ten feet without being roped in. That meant that I was stuck a hundred and eighty feet in the air without any way to get down should anything happen. I knew that if they cut the tree, as they repeatedly threatened to do, I would wind up going down with it. And that would be that.

Almond had taught me that a tree-sit is about more than just protecting a tree; it's a form of public outreach. That was one of the reasons we had persisted in asking for a cell phone, which we finally scored the day we had left for the tree. Stranded on the platform in the midst of all the chaos, unable to maneuver around the tree, I started dialing. I didn't know what else to do.

Almond had brought up a list of six different radio stations and newspaper numbers, so that's where I started. After several attempts, I got through to Annie Esposito's answering machine at community radio KZYX. I must have sounded like a psychopathic blathering idiot as I left my message.

It's hard to believe I ever shared this tiny platform with others.

"This is a forest defender and I'm up in a tree and they're cutting down the trees around me and they're climbing up and they are going to cut the line and I don't have a harness and there's activists in danger and I'm not sure what to do and . . . "

I was just babbling to the machine. I thought I'd hung up, but when I tried to make my next call, Annie was on the line.

"You just called me," she said. "What's going on?"

I tried to explain the insanity of the situation.

"Are you going to be all right?" she asked.

"I don't know," I answered truthfully. "I'm a hundred eighty feet in the air, and I'm stuck."

FIVE

EMBODYING LOVE

The onslaught continued for twelve days. In an effort to terrorize us into coming down, Pacific Lumber's crew continued to cut dozens of trees around us, not just the two growing off Luna's trunk.

"Look, we're going through the trees, we're going through the trees," they taunted. "We're going through the big trunk next."

To make a tree fall in a certain direction, they drive a wedge into it. Since I was raised in a Christian background, driving that wedge into the tree reminded me of the crucifixion. Jesus, an amazing prophet of love, was crucified by others driving spikes though him into a fallen tree.

Once the wedge pushes through that final crack and just a few threads remain, the tree starts to rip. As its final shreds are broken, it makes this horrible scream before crashing into those trees near it, breaking any branches in its way. It's a bone-shattering sound all the way down, and then *wham!,* the ground shakes and the air hums and everything vibrates with this fallen warrior.

Then for a single breath, everything gets deathly quiet. I didn't recognize that brief moment of silence in the beginning; it took going through the cutting day after day to hear that space when everything falls silent as if paying due respect. Then a split second later, the noise returns. With the bigger trees the loggers whoop and holler, as if they had killed a deer with big antlers.

About a week and a half into our sit, the loggers turned their chain saws on a big Douglas fir just west of Luna. They tend to cut the big ones uphill so they don't shatter upon impact, an all-too-frequent occurrence, because of high winds, that renders all the timber useless. That meant that this Douglas fir was being cut directly toward us.

One look told me that it would come close. I wanted to show people how recklessly these people were cutting—and how they were losing control of these trees—so I grabbed the video camera. Since loggers are paid per board feet instead of by the hour, they are pushed to cut bigger, better, and faster. So they work in conditions they shouldn't, and they take chances. But by choosing to cut such a big tree, so close to Luna, and in the day's high wind conditions, they chose to gamble with our lives. They—and I—nearly lost.

BY NOW, I had gotten comfortable enough to do some climbing without a harness. Ironically, I preferred it that way. This time, though encumbered by my video camera, I had ventured pretty far out on a branch, determined to get the shot. The famous camera, however, didn't want to focus. As I tried to get it to work, I heard them slamming the wedge into our huge neighbor, *clank, clink, clank, clink*. I knew this was the final push before the tree went smashing into the ground. *Creaaaaaak, groan, snap, crash!* It started to fall while I was still trying to focus. As it thundered downward, the Douglas fir hit the outer branches of Luna.

Wham!

I dropped the camera and slid about two or three feet, until I managed to reach out and grab onto a nearby branch. With my legs tightened around one branch and my arms wrapped around another, I hung sideways over a hundred feet in the air.

"Oh, s..t!" I heard the logger yell. Those are words you don't want to hear from your surgeon, your haircutter, or your logger.

"You almost killed me, and that's all you have to say?" I yelled back.

Holding on as tightly as possible, I finally managed to regain my balance and pull myself back up. Only once I was safe did the reality that I'd almost dropped over a hundred feet hit me.

"I'm out in the middle of nowhere, surrounded by angry loggers with big chain saws, the winds are intense, and I'm in a tree," I finally acknowledged to myself. "If an accident happens this high up, I'm dead. Period."

That was a scary moment. But things were moving so quickly that I didn't focus on it very long. After a momentary shudder, I got back down to business. The loggers weren't stopping, so neither could we. We had to keep trying to get through to them.

They didn't stop until December 23. Instead of twelve days of Christmas, we had twelve days of chain saws, all day, every day. Each time a chain saw cut through those trees, I felt it cut through me as well. It was like watching my family being killed. And just as we lose a part of ourselves with the passing of a family member or friends, so did I lose a part of myself with each fallen tree.

My first reaction was to want to strike out like an animal that's hurt or afraid. I wanted to stop the violence, I wanted to stop the pain, I wanted to stop the suffering, I wanted to stop these men who were cutting this hillside in complete disregard for the forest and the people's lives in Stafford below. I had hate for everything. I even had hate for myself because I was disgusted that I was part of a race of people with such a lack of respect. I was angry with myself for having been part of such a society for so long.

But then I began to pray.

I knew that if I didn't find a way to deal with my anger and hate, they would overwhelm me and I would be swallowed up in the fear, sadness, and frustration. I knew that to hate and strike out was to be a part of the same violence I was trying to stop. And so I prayed.

"Please, Universal Spirit, please help me find a way to deal with this, because if I don't, it's going to consume me."

You see that a lot in activists. The intense negative forces that are oppressing and destroying the Earth wind up overcoming many of them. They get so absorbed by the hate and the anger that they become hollow. I knew I didn't want to go there. Instead, my hate had to turn to love—unconditional agapé, love.

One day, through my prayers, an overwhelming amount of love started flowing into me, filling up the dark hole that threatened to consume me. I suddenly realized that what I was feeling was the love of the Earth, the love of Creation. Every day we, as a species, do so much to destroy Creation's ability to give us life. But that Creation continues to do everything in its power to give us life anyway. And that's true love.

If that beautiful source of Creation of which we're all a part could do that for us, I reasoned, then I had to find it within myself to have that feeling of unconditional love not only for the Earth as a planet, but also for humanity—even for those destroying the gift of life right in front of me.

But how to embody that philosophy? Somehow, I had to reach out on a personal level to the men cutting down the forest around me.

First, I tried reasoning with the loggers.

"What's the point in cutting at the rate you're cutting?" I asked. "Pretty soon all the trees will be gone. Then what will you do for work?"

I said whatever came to mind. I was emulating those activists who tried to talk issues with the loggers to slow them down. But I knew so little about the issues at that time, the loggers just blew me out of the water. They had been there, done that, heard it a billion times, and had an answer for everything.

So I started talking more on a human level. I always knew when I hit a good point, because they would immediately start up their chain saw or tell me, "F... you." Those were their standard responses when they weren't sure how to respond. I tried to get one guy to understand that old growth has a purpose, that Creation wouldn't have made things to grow old if they weren't meant to grow old.

"They're gonna fall over and die anyway, so what's your problem?" he responded.

I explained that it's important for the trees to fall over—it's part of the cycle, they become food for the Earth. I told him that we humans are supposed to become food for the Earth when we die, but we encase ourselves in coffins and other ways that will keep us from being part of the cycle.

Uninterested, he started up his chain saw. While he cut down yet another tree, I tried to figure out a way to get through to him. Suddenly it hit me.

"Hey," I yelled once the chain saw grew silent.

"What!" he yelled back.

"You have any grandparents?"

"Yeah, so what?"

"Are they alive?"

"Yeah, what's it to ya?"

"Well, why don't we just kill them? They're just going to fall over and die anyway. So what good are they?"

"F... you!" he screamed, immediately pull-starting his saw.

I KNEW THAT if I continued to debate politics and science—and stayed in the mind instead of the heart and the spirit—it would always be about one side versus the other. We all understand love, however; we all understand respect, we all understand dignity, and we all understand compassion up to a certain point. But how could I convince the loggers to transfer those feelings that they might have for a human being to the forest? And how could I get them to let go of their stereotypes of me? Because in their mind, I was a tree-hugging, granola-eating, dirty, dreadlocked hippie *environ-*

mentalist. They always managed to say this word with such disgust and disdain!

As I thought about this one afternoon, I remembered that I still had copies of one of the nicest snapshots ever taken of me, from my dear friend Kimberly's birthday just three months before I'd come up into the tree. I had sent them out to friends and family for Christmas, but I had a bunch left. Maybe seeing me made up and dressed in a silk suit and heels would shake up their stereotype of me!

Then I spied a canister in which we had stored some granola. I started laughing. I would show them that even "normal" people eat granola and that you didn't have to be a dirty, grungy, hairy, tree-hugging, leftist, extremist, radical *environmentalist* to like it. So I got a little plastic baggie, put in some granola and my photograph, sealed it off, and climbed down as low as I could. As usual, it was very cold and windy.

"Hey, buddy!" I said to the logger closest to the base of the tree. "I got something for ya."

"What?" he asked.

"I've got a gift for you. I want you to take a look at this. I've got a photograph of me, and I want you to see it, because I want you to see that I don't look like what you think I look like. Your preconceptions of me are wrong," I said, "and half the problem here is that you're dealing with me on the basis of your preconceptions. I'm trying to deal with you as a human being, and I'm not looking at you as some horrible logger. You won't do the same back, so I want you to see that I'm more human than you think I am. So, I'm dropping this thing down on the ground."

"You *better* not drop anything on me!"

"Chill out, man. It's really small, it's a plastic baggie," I retorted. "It's got some granola in it. I figure you're probably hungry, and it's really good stuff. And it's got a photograph in it. Just take a look."

So I dropped it down on the ground and stayed seated on my branch. Though I didn't have a real clear view, I watched him pick the package up. After a few moments, he yelled up.

"You're lying! That's not you!"

"That was taken on September 13 of this year. That's me," I assured him. "Granted, my tan has faded, my hair's not quite as clean and styled, I'm not wearing makeup, and I'm not dressed in a suit, but that's me. I want you to see how silly our preconceptions of each other are. We gotta let go of that and get over it. Get over our appearances."

"Damn! You really look like this?"

"Yeah."

"Then what the hell you doing up in a tree?"

That exchange definitely broke some barriers and took the edge off for a long time. The loggers joked that I had climbed into a tree simply because I hadn't found the right guy. Finding him for me, they decided, would solve all our problems. They were still angry that I was up in a tree and in their way, but they definitely started connecting with me on more of a human level. I was no longer this preconceived idea. As I became more real, they grew nicer. They never did tell me if they ate the granola.

＊

DURING THIS TIME, Almond, too, struggled with feelings of sadness and frustration. But whereas I experienced a transforming lesson of love and connection, Almond just seemed to sink deeper and deeper into despair and anger.

I love Almond a lot; I think he's a wonderful human being, and I'm really thankful for his friendship and insights. He taught me about the movement and about the history of activism while we were in the tree. Together we discussed strategy. He also impressed upon me the need for media outreach, which he saw as the biggest failure with the Luna tree-sit. The whole point of a tree-sit, he repeated, was to draw public attention to the problem. His energy and his view of the world, however, were a lot darker than mine. I am the eternal optimist. But, even so, it was really tough to be the eternal optimist considering everything else I was going through.

I felt like I was being hit with wave after wave of negative feelings, and I was having to struggle day in and day out to transform those negative feelings into positive ones. I began to be very concerned about Almond. Day by day, he seemed to grow more drawn. His already deep-set eyes and his striking cheekbones and thick eyebrows were a part of him when I met him, but as the onslaught wore on, his cheeks and eyes sunk in more and more. He started looking like a skeleton, his skin stretched tight over the bones underneath. His life seemed to be draining away in front of my eyes.

"Why are you still up here?" I finally asked. "You're sinking away, you're struggling. This is obviously not a healthy place for you to be. You can't keep going like this. It's just too much."

Thankfully, Almond recognized that he was in a downward spiral that would be hard to shake in the tree. He needed to go somewhere safe, without the intensity of the wind and loggers.

"But I don't want you to feel like I'm abandoning you," he announced. "I'm going to stick around. I'm going to try to help as much as I can."

Up to this time, Almond and I had been in this together. We had climbed up together and made Luna home. When we arrived in December, we had realized that the platform needed a lot of work. Basically it was falling apart. First, Almond set about fixing it up for the cold and rain. Then he built a better water-catching system. The system I had devised, which was a tarp bowl out in the open, worked really well until the wind picked up and blew all the water out. So Almond created a better one, more protected and accessible from the inside, which allowed us to collect most, if not all, of the water we needed from our tarps.

Almond's choice for our platform renovation was duct tape. I would go out after the storms and gather branches that had broken off Luna during storms and haul them back up to the platform. Then he would use whatever rope, twine, or yarn we had up in the tree, along with his ever-present duct tape, to make a frame to hold the tarps up. By the time he was done, our whole fort was one big web of silver tape!

An unexpected two-week dry spell graced our construction phase. The wind, however, was a constant. It never stopped. Then came Christmas Eve and Christmas Day, and for once it was wonderfully quiet. The chain saws and the loggers had taken a rest from ripping flesh from bone, trees from earth. The howling wind had slowed down as well. That was the best Christmas present Almond and I could have been given—the gift of peace! Though I didn't know it, I'd need those days of rejuvenation for what lay ahead.

Once our fort was resurrected, I began to settle into my new life. Certain basic details, like going to the bathroom, had almost become natural. *Almost* being the operative word.

After my first sit in Luna, I realized that this tree-sit hadn't been geared toward women. So I designed a more feminine toilet, consisting of a funnel with a hose. I changed this to a jar after wind kept ripping the funnel and hose.

I actually had two bathrooms. The first one consisted of the jar, which just got dumped overboard with each use. The wind would spread the urine over quite a large area in the time it took to fall a hundred and eighty feet. If Luna had been located in a dry forest, I wouldn't have been able to do that. The urine's acidity would have killed everything. But all the water in this forest made it okay.

The other bathroom was composed of a bucket lined with a heavy-duty trash bag. Nature had provided a storage space by striking Luna with lightning and burning out a huge cave. The bags were stored there until somebody offered to pack out the waste.

Cleaning up with our carefully rationed water also required a certain amount of ingenuity. My sponge baths were usually more like sponge dashes because of the cold. Once I had some privacy, I would heat water with a natural soap, strip from the waist down, and scrub like mad for all of about two minutes, if that, then grab a towel and dry off quickly before throwing my clothes back on. The temperatures and water and fuel rations just didn't allow for rinsing! Then I'd strip from the waist up and repeat the process. I washed my hair rarely, if at all. It took too much water and was usually so cold that if I got my head wet, I'd wind up sick.

In short, it wasn't easy. Whenever I'd get frustrated with these hygiene arrangements, however, I'd remember the seven families in Stafford who didn't even have homes, let alone a bathroom. That helped me readjust my focus back to what was important.

On the nesting front, I did what I could to make our small, wind-buffeted platform as cozy as possible. I added ambience and light by turning an extra little pot into a candleholder, which also doubled as a windscreen. For reading or writing during sleepless nights, I converted Almond's headlamp into a lantern. I had this only for the time Almond was in the tree. The rest of the time, I had only candles.

It made me realize how little I really needed to get by and how much growing up poor helped on that front. I knew what it was like to eat oatmeal every day for breakfast for a month because that was all we had. But in Luna, I learned that there are things a lot more important than a gourmet meal—or even a real roof over one's head.

Actually, not having a solid roof or walls helped me notice details that otherwise would have completely escaped me. When the rain started up again following our short reprieve, for example, I noticed its musky scent, like the sweet sweat of the Earth. And I heard two different noises as it fell—the rain dripping through the branches, and the rain hitting the forest floor in the clearing below. Raindrops falling onto my tarps sounded like popcorn popping. It made me think of my father, who loves popcorn, and my childhood because we used to have it just about every night at home.

Living in Luna made me start to pay attention to things so much more. Everything in life is its own little world, but most of us have gotten so caught up in our narrow arenas that we've forgotten to realize the magic and the beauty that are in all the other little interconnected worlds, too. The magical world of Luna is just phenomenal, down to how these trees disperse the water that falls from the sky.

I was sitting in the fog one day, unable to see past Luna's branches, when I noticed that the needles at the top part of the tree are knobbier than the needles lower down. Up high they look like gnarled fingers raking in the moisture from the fog and the rain. The water, drip by drip, gathers until it starts swirling down the trunk to the ground, over the smooth bark at the tip of the tree toward the increasingly shaggy bark down below, which absorbs more and more of the meandering flow. Toward the bottom of the tree, the needles become flatter and smoother. I imagine that's because they don't need to gather as much moisture. Instead, they act like a sprinkler system for the forest floor.

SOON IT WAS New Year's Eve, and the first anniversary of the Stafford slide. From up in Luna, Stafford looks like a miniature town. As I watched it from my perch on the hill, I was reminded of the place I visited in Chattanooga, Tennessee, as a little girl, with miniature trains, flying airplanes, hot-air balloons, and little towns with moving cars and people. Those memories, however, were quickly supplanted by the grim thought that a year after the slide, no material progress had been made toward protecting Stafford from another one. If anything, things had gotten worse.

"Okay," I told myself. "New Year's is supposed to be about resolutions. So what are mine?"

I thought about it. Resolution is about resolve. My resolution, I decided, was to take a stand like the redwood tree and not back down.

Even after they've been chopped into the ground, redwoods don't give up. Instead, they try to sprout new life. If they're not coated with herbicide (used after a clear-cut to control the vegetation so it doesn't compete for sunlight with newly planted farmed trees) or burned with diesel fuel and napalm (a common cleanup practice), they manage to regenerate.

The redwoods would be my guide. I would stay in Luna no matter what.

SIX

UNDER SIEGE

After twenty-five days in the tree, Almond descended to the
ground on January 4, 1998. I finally had the tree to myself for a
few days.

My reprieve, however, was a mixed blessing. Now that I
was alone, I had to deal with my fears by myself as well. Every
time I saw the guys from Pacific Lumber, my heart would start
racing.

"What are they going to do next to get me down?" I
wondered.

I knew Pacific Lumber wasn't going to just let me sit in
Luna. I was costing them money.

ON JANUARY 20, after days of rain, wind, and fog, the
weather finally cleared. The night before, I had heard it was
going to be nice, and immediately I assumed that the helicopters would be coming to lift out the fallen trees, which still
lay unharvested around Luna. An activist filmmaker named

Duff had called to say that he'd really like to come up in the tree and spend the night so that he could witness their arrival. Though I was in the mood to have some space, I agreed.

The next morning, I could see the Columbia helicopter logging crew hiking up the hill. The helicopter was indeed on its way. I'd overheard the loggers talking about intense updrafts, close to three hundred miles per hour. The longer I waited, the more my uncertainty and anticipation shifted to fear. I could feel my heart pounding.

"You'd better come down," advised a logger passing by the tree. "You coming down?"

"Nope," I answered, trying not to show how scared I was. "Staying up here, my friend."

"Then you'd better get ready for a bad hair day!" he shouted

I laughed. The tension had broken, and my heart slowed. But not for long.

Duff and I watched the fog begin to roll back out of the valley, knowing that as soon as that fog cleared, we'd be seeing the helicopter. In the distance, I could hear the rasping sounds of the walkie-talkies' static, along with garbled words I could not make out. The energy vibrated. Something big was going down.

Then the first sounds of helicopter, *thump thump thump thump thump thump,* getting louder and louder, *thump thump thump thump thump thump.* Duff and I looked at each other.

"Here it comes!" I said.

"Here it comes!" he agreed.

As if on cue, the helicopter rose from around the ridge. From far away, it didn't look so big. But then it got closer and closer, and the *thump thump thump* got louder and louder and louder, until you could actually feel the vibrations *thump thump thump* through the tree, as if she had been turned into a big drum. *Thump thump thump thump thump thump.*

That helicopter kept getting closer, and the closer it got, the bigger it got, until it got really, really big. It was heading straight for us.

"Oh, my God!" I muttered, looking at Duff.

I kept waiting for it to swerve, having seen the crew mark off the area around Luna with bright neon tape several days prior. I just knew the helicopter would veer around that tape any minute. It never did. Instead, it came straight for us, breaking the neon tape with its intense winds. I climbed out as far as I could on a branch to videotape it with a camera that had recently been sent up. I had barely locked my legs under the branch when the helicopter stopped right above my head.

Then the helicopter turned and stared me in the face. It was huge, the size of a passenger plane, with two propellers and cargo space, similar to the kind that the military uses to carry tanks and jeeps. I could see the silhouettes of the guys inside, with their helmets and headphones. I clung to my perch and tried to videotape.

It wasn't easy! Part of the time I held the camera up against my eye, but if that helicopter turned half an inch, the winds completely shifted. So the slightest movement would whip the branches—and me—first one direction, then another, threatening to tear the camera right out of my hands. My thighs tightened around the branch and my ankles locked together, I just gripped as hard as I could and tried to keep filming.

I felt like I was staring down the enemy. Here I was confronting this great big helicopter. At that moment, I hated that machine! The stench coming from its burning fuel was so bad that it nauseated me. The sound, like artillery rounds being fired right by my ears, or even between them, killed my head.

In the midst of it all, a strange sense of exhilaration invaded me. I wasn't being destroyed. I was holding on. I was catching this horrendous breach of aviation rules and common decency on film.

I felt like I was on the world's scariest roller-coaster—except that I was in a tree, and a twin propeller helicopter was creating the commotion. The adrenaline pulsed through my body. In some respects the moment seemed like it would never end, in others like it all happened in a split second. It was almost as if I had been caught in a cyclone; I spun round and round and round, and then *foomp!*, as quickly as the helicopter had come, it was gone. The fog was rolling back in, forcing it back to base, away from the logs it was supposed to pick up. Clamped onto the branch and reeling with shock, I watched it disappear out of sight.

It took me awhile to calm down. Once I was able to absorb what I had just gone through, I started shaking. When you're caught in the moment, the reality of what you're facing is not so apparent. Each split second becomes a moment of eternity that you just deal with as fast—and as well—as you can. When it's all done and you're left standing and alive, then it hits.

"Whoa! Whoa! Whoa!" I repeated over and over and over again once I was safe. Finally, after a long time, I looked back at Duff.

"Whoa!" I said one last time.

He looked at me, grinned, and shook his head.

"Whoa!" he echoed.

AT LAST OUR filming attempts in Luna had succeeded. After we taped the helicopter hovering right above my head, we sent it, on the advice of a fellow activist, to the Federal Aviation Administration. It's against regulations to fly within two hundred feet of humans, so sending the tape must have been effective because they stopped buzzing me shortly thereafter.

For once, Pacific Lumber seemed intimidated. They didn't return within that two hundred feet to pick up all those logs they had cut down by Luna, although they did continue to log the rest of their timber harvest plan. Hundreds of thousands of dollars in timber would sit there until I was out of the tree. They would wait for me to leave before they collected them.

They weren't the only ones who wanted me down. Even as the helicopters were taking the logs and trying to blow me away, many of the Earth First!ers decided that I should leave Luna. They'd never had a tree-sitter go through the winter, and they were afraid I might get injured, which would make the movement look bad. They also didn't feel like they had the resources to keep the tree-sit going. Then there were those who didn't like the fact that Luna had become this kind of rogue tree-sit that was defying all the rules and regulations.

Earth First! is a diverse group that operates under the rule of consensus. I didn't abide by that. I didn't ask anyone's permission to stay in Luna, I just did it. Ironically, their opposition only encouraged me to continue on.

When you get right down to it, my parents are partially to blame for my rebellious and stubborn character. They taught my brothers and me to question authority. My father preached, "Never take what someone else says as absolute truth. You have to question it and find out if it *is* the truth." Once we knew it to be the truth, my parents taught us to stick by that and never back down unless we were proven wrong. My father says I learned that lesson too well.

"You always questioned me and your mom, and you never agreed with us," he says. "Your truth was never the same as ours, and you were so darn stubborn about your truth."

So it was with Earth First! and Luna. Toward the end of January, an activist named Cat asked to come up and spend an evening. At that time, I pretty much said yes to anyone wanting to come into the tree, because I didn't feel like I had the right to ask people to leave me alone.

"Have you heard that Earth First! wants you to come down?" she asked during our initial conversation in the tree.

"Yeah," I replied.

"Did you know they're having a meeting tomorrow night to discuss how to get you down?"

I blinked.

"No, I didn't realize that," I confessed.

"Julia, do you consider yourself an Earth First!er?" she asked suddenly.

I looked at her and laughed. Then I launched into a tirade that had been brewing for a long time.

"How can I consider myself an Earth First!er when they didn't want me around in the first place? How can I consider myself a member of this group when I didn't find out until the second time I was up here that this was an Earth First! tree-sit? I didn't even know what Earth First! meant. How can I consider myself an Earth First!er when it's Earth First!ers who are trying to get me to come down? To be a part of something, you have to have an affinity with it, and I don't feel an allegiance to whatever the heck Earth First! is. So, no, I don't feel like I'm an Earth First!er."

"Well, don't you think that's a problem, considering the fact that you're in an Earth First! tree-sit?"

"I came into this tree-sit because they didn't have anyone else to sit. They didn't tell me they were Earth First! or that there were certain rules and regulations I needed to go through and agree to before I came up here. They needed someone, I answered the call, and here I am. I'm not trying to go against anyone, but I just can't see why you guys would let this tree come down. Why even have a tree-sit if all you're going to do is give up and come down? Especially while they're still logging in the area! The least you can do is wait until they're not around anymore. To come down now means that Luna would die. I can't see that happening. I can't see how you guys would even let that happen. I'm sorry, but I'm not Earth First!, and I'm not coming down. You go back to that circle, and you tell them that. You tell them they can take all their stuff, they just need to leave me one rope and one harness. They can take everything else. I'll use my money to get whatever's needed. I'll take care of it. They don't need to worry about their resources. They don't need to worry about Earth First! being made to look bad, because I'll make sure to let people know I'm not an Earth First!er. They can write this tree off, but I'm not letting it go. There's no way I'm letting this incredible tree fall. I'm not going to do it. As long as I have the ability to keep this tree standing, I'm staying up here."

I felt better after getting all that off my chest.

Even as a kid, I've always stood up for what I believed in. And it's cost me. When I was in the eleventh grade, some friends and I got attacked by teenage boys with bats in the parking lot of a fast-food restaurant. It turned out that I went to school with seven of them. Then I discovered that this wasn't their first offense: a lot of them had stacks of complaints against them, including a prior incident during which they had attacked a young guy at the fair with a crowbar and put him in the hospital with stitches and broken bones. Despite the severity of the incident, the young man was too scared to press charges, so no action had been taken against them.

"My God, I'm lucky that I'm alive," I thought. "They're going to kill somebody next."

So I pressed charges. That's when their girlfriends started in on me. It was too surreal to be true. They would surround me and slam me into lockers or corner me in the bathroom and shove me through the stall doors. But I prosecuted anyway, and some of the attackers were put away.

I hadn't backed down then, and I wasn't about to back down with Luna. If I had bowed to Earth First!'s wishes, Luna would have been cut in two seconds flat. I couldn't bear that.

Still, it was difficult to have people from within the movement try to get me to come down when I knew I should stay up. Cat climbed down the next day, leaving me feeling angry and disgusted. The thought of all these people sitting down there in a circle talking about me and determining that I didn't have a right to be in Luna, as if they owned the tree, made me crazy.

I talked to Shakespeare about it that very night when he came to visit.

"Shakes, I want to know how you feel about this tree-sit and about my being up here."

"Julia, I'm with you," he assured me. "If you want to stay, I'm with you. I'll help you any way I can."

I asked if he would be willing to hike in enough supplies for me to be self-sufficient should Earth First! decide to pull its support. He agreed willingly.

"I won't run off and leave you up here," he promised, looking me in the eyes. "Julia, you can count on me."

I knew he meant it.

⁓

THE NEXT MORNING changed everything.

As I climbed outside the shelter, I saw two men with bright yellow jackets and badges pull up on four-wheelers.

"Uh-oh!" I gasped, scrambling back onto the platform. "Shakespeare, something's going on. There are two guys out there with jackets, uniforms, and badges."

Swearing an oath, Shakespeare decided that one of us should climb down to take a look at what was going on, while the other got on the phone. Since I had more experience dealing with the media—which wasn't saying much—I opted to dial while he rappelled partway down.

"They're throwing lines up into the tree," he yelled up from his new vantage point.

"They're coming after us!" I thought fearfully.

The two guards, however, were both heavyset men who were in no shape to be climbing trees. *They* wouldn't be coming up into the tree after us, but they could be setting things up for somebody else. I couldn't stop thinking about what I would do if someone did come up after us.

"I'm committed to peaceful activism, but where is the line?" I asked myself. "What can I in all good conscience do to protect this tree?"

I wasn't just going to go limp and let them lower me down, that was for sure. I wasn't going to make it easy for them to get this tree down. But at the same time, I knew I wasn't going to try to beat up anyone who came up or endanger anyone's life. I simply don't believe there is justification for violence. So I did a lot of fast soul-searching. Climber Dan had climbed the other tree really quickly, so I knew that if they *were* coming up this tree, I didn't have a lot of time to figure out how to respond.

Controlling my fear, I called the two local radio stations I knew how to reach and told them what was going on. Telling them to stand by for more information, I climbed back down part of the way to check in with Shakespeare.

"Shakes, what's going on?" I yelled down.

"Well, I asked them, and they told me they're just going to starve us out."

"What?"

"That's what they said. They're just going to starve us out."

Within the hour, additional security guards showed up.

"Look, you have twenty-four hours to come down," announced one of the new arrivals. "If you come down by noon tomorrow, you won't be arrested. After that, it's free game. We *are* going to get you down. There won't be any more supplies coming to you guys. We're cutting them off. So you *will* come down. Then you'll be arrested."

My heart was racing like mad. These people had hovered a helicopter above our heads, they had cut a line with an activist on it. What would they try next? I was scared. We were so far away from anyone who could help, out in the middle of nowhere, surrounded by people who didn't care what happened to us, who just wanted us out of the way. Some of them would probably have been happy to see us die. I couldn't shake the feeling that they would hurt me and not think twice about it.

Shakespeare and I tried to play it cool. We took turns chatting, and even joking, with the guards. Finally, we figured out that they had just thrown lines into the tree to attach the tarps. I calmed down.

Once we'd made sure they weren't trying to scale Luna, we climbed back up to the top of the platform to talk strategy.

There was plenty of food and water in the tree for one, we agreed, but not enough for two. So we began to ration both carefully.

To further complicate matters, Shakespeare was supposed to be in court a few days later on previous misdemeanor charges including "failure to appear." If he went down and was arrested, as threatened, he'd get slammed for good. So he was stuck in the tree with me.

He considered climbing over to the one tree we were still attached to, since he'd be far enough from the spotlights there to rappel down, and take off at a run the moment he hit the ground. But our rope wasn't long enough; he would have been left hanging about forty feet in the air. Jumping that far with a backpack wasn't an option.

So he stayed, and our food became increasingly limited. I was good at rationing, but Shakespeare wasn't. His way of feeling okay was to eat. As I doled everything out bit by bit, Shakespeare eyeballed the stores and moped. He was really torn: should he stay or should he go? Then, after two days, one of the guards unwittingly weighed in.

"Why don't you guys just come down?" he said. "Even though the deadline passed, we promise we won't arrest you."

That gave me an idea.

"Let's tell them we'll send one person down and it'll be you, and that if you make it back to our group and call me to say that you made it safely, I'll consider coming down," I proposed.

So that's what he did. The guards even drove him all the way down the hill and dropped him off at somebody's house. After Shakespeare called to let me know he was okay, I broke the news that I wasn't coming down.

"But your buddy made it safe! You're a liar!" they screamed, enraged.

"I never told you I would come down. I told you I'd *consider* it. Well, I've considered it, and I can't imagine letting this tree fall. I'm sorry, I can't come down."

That really angered the guards, especially since some of the winter's coldest, wettest storms had moved into the area. They stepped up the attacks after that. I was a good target for them to vent on, and they vented hard. They blew bugles and air horns all night, keeping me awake for all but an hour or two. They trained floodlights on the tree, the generators going *grrrr*. They screamed at me, calling me dirty names that I don't even like to think back on. They threatened me.

"If you don't come down right now, when you do make it down, we're going to beat you into a pulp for making us sit out here in this bad weather," someone yelled. I couldn't tell if this was just an "idle threat" but I was not in a great position to figure that out.

Two to three guards worked each twelve-hour shift, rotating out at noon and midnight. As the days wore on and the storms worsened, the men got increasingly irritable and hostile. All except Kalani.

I met Kalani on the third or fourth day of the siege. I had leaned back on a branch and was singing a song called "Love in Any Language" in response to the steadily mounting stream of abuse. Kalani was up on the clearing that lies on the logging road right behind the tree-sit. I didn't realize he was watching at first. Then I saw him standing there. When I finished the song, I waved to him. He waved back.

"You know, you're a good singer," he said.

I started laughing.

"No, I'm not, but thanks anyway."

He laughed, too.

"No, really," he said. "I'm in a band, I play drums. You need to sing with us. You need to come down out of that tree and come sing in my band with me."

"Nice try," I said, and we both laughed again.

Kalani and I struck up a dialogue over the subsequent days. We connected. We talked a lot about who we were, about our lives. I really tried to honor him as a human being and take an interest in who he was as a person. I could tell he was having a rough time in his life.

"I feel like there's something wrong with your marriage, because you've mentioned your dog a lot more times, and with a lot stronger energy in your voice, than you have your wife," I told him at one point.

Kalani's presence kept me from going insane during that period. I was so afraid of what those other men were going to do to me, but every other twelve-hour shift, Kalani would be on duty. If something went down while he was there, I knew I wouldn't be beaten or raped or whatever else the others wanted to do to me.

"You know, if you'll come down, I'll buy you dinner. I don't know much about you vegetarians, but I'll take you wherever you want to go and we can talk," he cajoled at one point. "I'll take you to hear my band play. Just come down. You've done a good job, you've made your point."

That was what they all liked to say, that I'd made my point. I *had* made points by being in Luna, but the tree was still not saved. So my job wasn't done. And I sure wasn't going to come down because they were trying to starve me down. That made me all the more determined to stay up.

"You know, Kalani, I would love to have dinner with you

someday. I think that would be great," I said. "But I'm not coming down, friend, I'm sorry. I can't do it. I can't see this beautiful tree fall."

"Well, you've done a really good job, but sooner or later they're going to get this tree," he insisted. "So you can't sit in it forever. Why not come down now, where you'll be safe? I'll promise I'll watch out for you, I won't let them do anything."

His buddies cursed him for being friendly to me.

"You're not going to get her down that way!" they screamed.

"You think you're going to get her down your way?" he answered. "You think abusing her is going to bring her down?"

He was right. If anybody was going to bring me down, it was Kalani. At least I felt safe with him.

The winter storms really started to hit then, making life even more trying. The temperatures plummeted. As much as I feared and resented the security guards, I felt bad for them. Charles Hurwitz, the head of Maxxam Corporation, should have been the one sitting under the tree in such nasty weather. I prayed that they would get transferred back to their warm trailers.

I prayed for my own comfort as well, because I was getting seriously cold. All I had in the way of cold-weather gear was a wool sweater that had been given to me the day I left base camp, one T-shirt, one thermal shirt, one pair of thermal pants, and a pair of wool pants, which saved my life. Even my sleeping bag was a lightweight synthetic not made for really low temperatures.

It was hard enough to handle being cold, but being wet and cold nearly did me in. I ended up with frostbite. It didn't happen all at once. I just couldn't get good circulation. To get my blood pumping, I had to climb around, but the hail, sleet, and fierce, cold winds drove me inside my tarps. The few times I tried climbing outside, I felt as if I was exercising in an industrial freezer with the fan cranked up high.

Now when I climb around Luna, I do so in bare feet. I had to abandon my shoes early on in the sit. I couldn't stand the feeling of separation from the tree. With all that stuff between my foot and the branches, I couldn't tell if what I was about to stand on was strong enough to hold me or if my foot was on the branch securely. I couldn't feel Luna's life force or take instruction from her about how to climb. So I took off my shoes after the first few days and hung them from a branch, where they have stayed.

But barefoot in winter in a tree on a hill in the freezing wind is not always a great idea. Just before the helicopter siege, my feet had begun to hurt. They hurt all the time. At first I didn't know what was wrong. I would rub them, to no avail. Then they started turning colors.

"My toes are going from red to white to blue," I joked, using humor, as usual, to help me through. "They've become patriotic."

Finally, one of the activists who had come to visit told me I had frostbite, and sure enough, the patriotic colors turned black and purple.

It reached a point where the pain was excruciating. It hurt so badly that I would have to stuff clothes inside my sleeping bag to keep the bag from touching the edges or tops of my feet. Figuring that the heating action of Tiger Balm might help warm them up, I dabbed on a coat as lightly as possible. I couldn't even rub it in because the pressure would have been too painful. Then I wrapped my feet in gauze. Thankfully, they eventually healed.

I also managed to break the little toe in my right foot at that time. I was climbing around in the cold, and I always stub my toes when I climb around in the cold. I knocked it good that day, but my foot was so cold I didn't realize I had broken it. When my feet started to warm up, however, the pain was agonizing. I looked down at my toe and saw a huge purple and black line running down the side. I got the Tiger Balm back out, then wrapped toilet paper around the toe because I had run out of gauze. I duct-taped a piece of cardboard to my toe and taped my makeshift splint to the other toes. With enough duct tape, you can fix anything.

I was starting to feel like my whole life was held together by duct tape, however. I just hoped it would hold.

"Insanity of Windstorms
    High in the Sky"

I drown
deep into the vast unending
bottomless
pool of thoughts
patterns
swirling
spirals, dots, zig-zags
on and on into eternity
slow down
I am sick
from the whirling, twirling
spinning on and on
give me rest
peace
stillness in the quiet
of the night
smooth surface of a lake
not a ripple
to distort the perfect moon
my tortured mind seeks
comfort
respite
solace
solitude   '98

SEVEN

THE STORM

Despite Kalani's help on the emotional front, the siege was taking its toll. For starters, I was running out of battery power for communication, so we had to pull off a resupply. We tried in the middle of the night. The attempt totally failed. The security guards, who stayed on duty round the clock, kept the floodlights illuminated. There was a demonstration staged, I later found out, down on Highway 101, which I can see from the top of Luna. A group of people was trying to create a diversion so the resupply could get to me. I found out about this later when I used my new cell phone batteries to call Almond.

The guards had become very vigilant about stopping resupply efforts. "Nobody's going to get to you," they said when three people were caught. "You are not going to get resupplied. We are going to starve you out, so you might as well come down."

One of the three people, a seventy-something woman known as Grandma Rosemary, told them she was my grandmother and that she wanted to hike up the hill to tell me to come down. The security guards told me they couldn't let my grandmother hike up for fear she'd have a heart attack; I had to come down to see her. If only they had known! This amazing woman had hiked the hill three times and later climbed all the way up the tree!

"I don't know who's down there, but it's not my grandmother," I retorted.

Then I asked them to describe the other two people they had stopped. One of the descriptions fit Almond.

"Look, if you want me to come down, find out if one of the people you have is Doug Fir," the media name Almond was using at the time. "If he's there, bring him up here. If he tells me to come down, I'll come down."

I actually hoped in a way that they did have Almond and that he would ask me to come down. I was so sad thinking of losing Luna to these destructive, greedy men. But I just couldn't imagine facing another freezing cold, stormy night like the preceding one. The night before had been one of the scariest nights of my life.

In Luna, the wind is a constant. And wind does something to you, something that rain doesn't. It makes your thoughts go wild. You can't focus. You can't read or write or paint or think. You feel disconnected and ungrounded. The sound of tarps whipping in the wind drives you crazy. You just sit here with a glaze, while your mind gets pummeled.

It was January, and gale-force winds, along with rain, sleet, and hail, had set in. I grew up with storms. I knew they passed.

These didn't. This, as I later found out, was El Niño, one of the worst winters in recorded history in northern California.

The storms had been growing in power every week, each day worse than the last. The night before, the storm had completely shredded my tarps. I sat wrapped up like a burrito in what remained of a single tarp, getting beaten with hail and freezing. Another picked up the platform and tossed it about like a matchstick. That scared me so badly, I gave myself into the craziness of it all and laughed hysterically, holding onto my rapidly slipping shreds of sanity. I prayed like I haven't prayed in a long, long time. I prayed to every power and to every God. I begged and pleaded for my life, the life of my friend next to me, and the life of the goddess in whose arms I was being held. I even said a prayer for the safety of the Pacific Lumber security men on the ground below us.

The storms were so loud and I was so cold so much of the time that I couldn't sleep. After six or seven of these sleepless nights, I started to break down. Suddenly, I couldn't stop crying. The intensity I had undergone over the preceding weeks had drained everything out of me. But without a decent night's rest, I was never able to recharge. I reached empty, and still I kept draining.

"I cannot survive any longer this way," I thought. "I cannot go one more night on no sleep, just hearing the howling of the wind and the sleet pelting me through the cracks. I can't do it anymore. I just can't."

But the storms would get even worse—a lot worse. It was in

this weather that a group of Earth First!ers and others tried to resupply me with food and batteries.

The guards held the line. "I'm not bringing any more of you Earth First!ers up this hill!" screamed one of them. "It's trespassing! It's private property! Just get the hell off it!"

By this stage, nine or ten Pacific Lumber people had gathered at the base of the tree, including one of their top brass. Despite their numbers, I kept clinging to the notion that somehow a new supply run would get through. I stared down the hill through intermittent hailstorms all afternoon. Nothing. I began to cry out of sheer exhaustion.

Finally, I was forced to accept that the supply runs had failed. I was so broken by that point that if anybody I trusted from my side had told me to come down, I probably would have.

Late that same afternoon, I saw these little colored specks coming up the mountain. In my sleep-deprived state, I still assumed that the supply run had failed again.

Then my pager sounded, and I saw our prearranged signal indicating a resupply. We had set up a few different codes via pager so I would know when a resupply team was headed out, when it was nearing the tree, and when to drop the duffel bag we used to haul up equipment and supplies. The code told me they were near.

I had to create a distraction. Several days before the siege, the activists had sent up a banner that said "Earth Jobs First!" So I unfurled all thirty-five feet of it—*Whooom!*

"Heads up!" I yelled to the guards, afraid that the banner

might break some branches. Meanwhile, the resupply group of twenty activists crept stealthily closer. To cover up any noises, like breaking twigs, I started singing the same song I'd sung to the guards, and the loggers before them, as loudly as I could.

> *Love in any language,*
> *Straight from the heart,*
> *Pulls us all together,*
> *Never apart.*

The security guards didn't think anything about it because I'd been singing that one song over and over the whole time. Suddenly Shakespeare walked up to them.

"Hey, guys, how's it going?" he asked casually.

Nineteen other people popped up from the surrounding greenery and yelled at the tops of their lungs, "Twenty-three!" That was how old I was at the time. It was also our code to drop the haul line. I had previously lowered it part of the way through the branches. When I heard the code, I dropped it the rest of the way down.

The security guards and the activists, the latter cracking jokes and running around in circles, all dashed to the rope. Many of the activists had stuff sacks in their hands, but only some of those had supplies in them. The security guards didn't know whom to tackle. All they could do was go for the rope. Before they got there, one of the activists managed to clip a bag onto the rope.

"Twenty-three!" they shouted again, my cue to pull the haul bag back up.

A guard grabbed for it, but the activist took a flying leap and hit the bag downhill. The hill Luna stands on drops so steeply that one side of the tree's base is twelve feet lower than the other. By hitting the haul bag down toward the ravine, it flew above the security guard's hands. That was all I needed. My first yank pulled it up well above them, and then I was able to haul it safely to the top.

That same maneuver worked once more. The second time the haul bag was in a security guard's hands, but he didn't have a grip on it yet, so an activist jumped through the air and hit it again, sending it flying. While I pulled it up into the tree, the activists posed for pictures with big grins on their faces and then took off back into the bushes before the security guards showed up again.

As I unpacked my fresh fruit, cell phone batteries, and propane with which to cook tea on my stove, I thought about those twenty people—some Earth First!ers, others from all walks of life—who were willing and ready to be arrested that day, and all the people who staged the demonstration below in order to resupply me. The beauty of so much love gave me the burst of energy I so desperately needed. Suddenly I felt supported instead of all alone. I was recharged and ready to keep going.

The successful resupply turned many Earth First!ers around as well. In their eyes, this was war! The siege had provided them with a chance to shake things up, to be in the face of the corporate structure, which they love to do. That's one of their ways to publicize an issue. Now they were rallied behind the Luna tree-sit.

After the resupply, Carl Anderson, Pacific Lumber's head of security, chewed the guards up one side and down the other.

"What is wrong with you guys?" he demanded, cussing them out.

"Man, they were outnumbered," I called down. "There was nothing they could do."

Of course, without *my* team, there would have been nothing I could have done. Together, we were the tools trying to keep this vehicle of a movement running. In a car race, somebody is at the steering wheel, and at the finish line that driver gets the trophy and the spotlight. But everybody knows that he had a top-of-the-line car and a top-of-the-line pit crew, that all the bolts were tightened to perfection, and that everything was lubricated just right. So the trophy is not just for that car's driver, it represents the whole effort of all the people who worked on the car. And so with this movement: a victory comes from the efforts of everyone, not just me. Our trophy will be to find permanent protection for all the old growth while we save Luna. And we will all share that prize.

⁓

THE STORMS GOT so bad that the guards pulled out two days after the resupply. Every night, the wind ripped huge branches off trees and flung them to the ground, making it extremely dangerous for anyone on the forest floor. Plus, the guards knew that the activists would continue to outwit them unless they brought in a whole bunch of extra security, and the company didn't seem willing to do that. So finally, two days after the resupply, they left without a word.

I never saw Kalani again, though I spoke with him over the telephone a couple of times, finding out in the process that he had quit Pacific Lumber security. I think of him as an angel, because angels are the people who pop into our lives at a needed time and then disappear without a trace, having helped us through a critical point.

⌒~~~~⁄

ON DAY SEVENTY-ONE of the tree-sit, a photographer hiked up to Luna with a group of people and asked if he could come up and take photos. I hesitated. But he knew how to climb, and the group had brought all this food up for me. Feeling obligated, I agreed. Eric Slomanson turned out to be an upbeat, funny man who quickly put me at ease. Over the next two days, he took pictures while we talked and clowned around.

"You know, dahling," he said, pretending to be an agent, "if you really want to do this right, you've got to stay until the hundred-day mark, because the world record is ninety days, and Americans love record breaking, anniversaries, and numbers. So not only will you break the world record, you'll have a nice round number of a hundred."

We both laughed, but he was serious about my staying up.

"Are you crazy? That's three weeks away!" I exclaimed. "It's just not possible. Look at me. I'm dying here!"

"Yes, it is! You've been up here this long. Three weeks is nothing to you."

"There's no way I can last another three weeks," I insisted. "There's just no way."

I was so tired, cold, and wet, the only thing I could think about was a hot shower and a bunch of bottles of wine in a row. My nerves were shot. Every noise I heard convinced me that the tree was under attack again. But the quiet also scared me, because then I didn't know what was going on.

I even found myself missing the security guards. I had gotten used to their generators, their lights, their dogs, their chaos, and their cussing. And then all that went away, leaving me to wonder what the next barrage would be. It had to be something. Surely, they couldn't just roll over; they would have to save face.

Losing my mind from a lack of sleep, food, positive results, and emotional support, I began to feel like my whole being was under attack. I was near the breaking point, unable to fend off the devastating impact of the elements. When I'd get wet, the chill would work itself into my very core. My shivers would wrack my body for hours, even after I had dried off and settled into my sleeping bag.

"I'm soaking wet, I'm cold, and I'm miserable," I confessed to myself.

The cold and damp aggravated all the physical ailments that still lingered from my wreck. My shoulder bothered me. So did my back and neck.

"Man, why am I even here?" I wondered.

NOT LONG AFTER the security guards left, I thought that a bit of music might calm me down—anything other than the incessant flapping of tarps in the wind. Because of the helicopters and the storms and the siege, I hadn't listened to my radio in weeks. So I pulled it out of the five-gallon storage bucket and started messing with it. Finally, I got it to work.

I put on the headphones and nestled down into my sleeping bag, ready to relax. Instead of tunes, however, I heard warnings about another storm watch and upcoming seventy-mile-an-hour winds.

"That's for down there," I thought with trepidation. "God only knows what it's going to be like up here."

The mere thought terrified me. I kept wondering if this was a sign that I should go down. I remembered my father preaching about a man who hears that a flood is coming. So he goes and prays to God.

"God, please protect me," he asks. "Keep me safe from this flood."

"Ask and you shall receive," God responds.

The floodwaters begin to rise. As they flow over the front stoop and enter the first floor of his home, some people come by, paddling a canoe.

"Hey, brother, would you like a ride to safe ground?" they ask.

"No, no, that's okay," says the man. "I prayed to God, and he said he's going to take care of me."

The floodwaters continue to rise, and the man has to move up into the second story of his house. More people come by, this time in a motorboat.

"Hey, we're here to save you!"

"Oh, no, that's okay," says the man. "God said he is going to protect me and take care of me."

The floodwaters continue to rise, and he's standing on the roof of his house, and some people fly by in a helicopter.

"We're here to save you! We're here to save you!" they shout, hovering over him.

"No, thank you, God said he's going to protect me," the man responds.

"Are you nuts? Look at you, man! Your house is under water. We're the only ones who can save you. Hop on!"

"No, no, really," says the man. "God will save me. God said he would save me, so I know God will."

"You're nuts," they say as they fly off.

Sure enough, the man drowns.

He stands before God in heaven. "God, your answer to my prayer was, 'Ask and you shall receive.' Why did you let me drown?" the man asks.

"I sent a canoe, a motorboat, and a helicopter to save you!" God retorts. "What more do you want?"

⌒

I DIDN'T WANT to let Luna fall. But I didn't want to be stupid. I didn't want to miss my opportunities to make it out alive. I wondered whether the worsening storms had been my sign—and my help—from up above to get out of the tree.

"Maybe this radio broadcast is going to be the last helper that God is going to send my way," I thought. "Or maybe my fear is simply trying to put me under its power and manipulate me into going down to where it seems safer, where I don't risk falling a hundred eighty feet and smashing into the ground, with branches skewering me like somebody's barbecue."

I couldn't tell. The only thing I knew for sure was that I wished I hadn't heard that radio alert. I tried meditating. I tried praying. I still didn't know what to do. I didn't want to get blown over or get blown out of the tree. I wasn't ready to die yet. If it was my time, then it was my time. But I didn't need to bring my time upon myself before it was meant to be.

I tried to figure out what I was supposed to do. I didn't feel any need to play Superwoman, but I knew that I had given my word that I wasn't coming down until I had done everything I possibly could to protect this area. To come down because I was afraid of a storm would be to break my word, and I believe beyond a shadow of a doubt that we are only as good as our word. If our actions don't meet our words, our value as people is lessened. That's just the way I was raised.

Still, I was torn. My survival instinct was telling me to go down to the ground, that the Pacific Lumber people had all left, and that I could just climb back up in the morning and nobody would ever know. But that would mean breaking my word, and I just couldn't do it.

Before the storm ended, however, a promise would be the last thing on my mind. I would just be trying to stay alive— and not doing a very good job of it.

⌒⌒⌒

THE MOMENT THE storm hit, I couldn't have climbed down if I had wanted to. To climb you have to be able to move, and my hands were frozen. Massive amounts of rain, sleet, and hail mixed together, and the winds blew so hard I might have been ripped off a branch.

The storm was every bit as strong as they said it would be. Actually, up here, it was even stronger. When a gust would come through, it would flip the platform up into the air, bucking me all over the place.

"Boy! Whoaaah! Ooh! Whoa!"

The gust rolled me all the way up to the hammock. Only the rope that cuts an angle underneath it prevented me from slipping through the gap in the platform.

"I'm really ready for this storm to chill out. I'm duly impressed," I decided. "I've bowed and cowered once again before the great almighty gods of wind and rain and storm. I've paid my respects—and my dues—and I'd appreciate it if they got the heck out of here."

My thoughts seemed to anger the storm spirits.

"Whoa! Whoa!" I cried, as the raging wind flung my platform, straining the ropes that attached it.

"This is getting really intense! Oh, my God! Oh, my God!

Okay, never mind, I take it back. Whoaaah!"

The biggest gust threw me close to three feet. I grabbed onto the branch of Luna that comes through the middle of the platform, and I prayed.

"I want to be strong for you, Luna. I want to be strong for the forest. I don't want to die, because I want to help make a difference. I want to be strong for the movement, but I can't even be strong for myself."

It seemed like it took all my will to stay alive. I was trying to hold onto life so hard that my teeth were clenched, my jaws were clenched, my muscles were clenched, my fists were clenched, everything in my body was clenched completely and totally tight.

I knew I was going to die.

The wind howled. It sounded like wild banshees, *rrahhh,* while the tarps added to the crazy cacophony of noise, *flap, flap, flap, bap, bap, flap, bap!* Had I remained tensed for the sixteen hours that the storm raged, I would have snapped. Instead, I grabbed onto Luna, hugging the branch that comes up through the platform, and prayed to her.

"I don't know what's happening here. I don't want to go down, because I made a pact with you. But I can't be strong now. I'm frightened out of my mind, Luna, I'm losing it. I'm going crazy!"

Maybe I was, maybe I wasn't, but in that moment I heard the voice of Luna speak to me.

"Julia, think of the trees in the storm."

And as I started to picture the trees in the storm, the answer began to dawn on me.

"The trees in the storm don't try to stand up straight and tall and erect. They allow themselves to bend and be blown with the wind. They understand the power of letting go," continued the voice. "Those trees and those branches that try too hard to stand up strong and straight are the ones that break. Now is not the time for you to be strong, Julia, or you, too, will break. Learn the power of the trees. Let it flow. Let it go. That is the way you are going to make it through this storm. And that is the way to make it through the storms of life."

I suddenly understood. So as I was getting chunked all over by the wind, tossed left and right, I just let it go. I let my muscles go. I let my jaw unlock. I let the wind blow and the craziness flow. I bent and flailed with it, just like the trees, which flail in the wind. I howled. I laughed. I whooped and cried and screamed and raged. I hollered and I jibbered and I jabbered. Whatever came through me, I just let it go.

"When my time comes, I'm going to die grinning," I yelled.

Everything around me was being ripped apart. My sanity felt like it was slipping through my fingers like a runaway rope. And I gave in.

"Fine. Take it. Take my life. Take my sanity. Take it all."

Once the storm ended, I realized that by letting go of all attachments, including my attachment to self, people no longer had any power over me. They could take my life if they felt the need, but I was no longer going to live my life out of fear, the way too many people do, jolted by our disconnected society. I was going to live my life guided from the higher source, the Creation source.

I couldn't have realized any of this without having been broken emotionally and spiritually and mentally and physically. I had to be pummeled by humankind. I had to be pummeled by Mother Nature. I had to be broken until I saw no hope, until I went crazy, until I finally let go. Only then could I be rebuilt; only then could I be filled back up with who I am meant to be. Only then could I become my higher self.

That's the message of the butterfly. I had come through darkness and storms and had been transformed. I was living proof of the power of metamorphosis.

"LOUD QUIET"

Eerie
Quiet Stillness
this calm after the storm
The air is pregnant with the weight of nothingness
yet something just the same
I'm jumpy
My nerves are shot
My mind creating noises
because it can't handle such dense oblivion
I'm poised waiting to hear the rush
my senses being overwhelmed
I feel like a rubber band
stretched to its last possible point of tension
before the world I know splits
into a crazy reverberating apocalypse of chaos
I listen hearing nothing but the occasional rumbling
of the highway far off in the distance
yet my dillusional mind recreates last night's onslaught
over and over again within me
As if I could ever forget ❧ ∾98

With the butterfly that landed on my finger while on a hike in 1980.

EIGHT

REGENERATION

When I was six or seven years old, a butterfly landed on me and stayed with me for hours while I hiked in the mountains of Pennsylvania. Since then, butterflies have always come to me during times of need, sometimes in reality and other times in visions and dreams. At one point, when I was feeling extremely despondent, a vision came to me of a butterfly poking out of a cocoon. When it finally broke free it was a magical butterfly with prismatic colors. As the butterfly emerged, the cocoon's brown shell turned into a shimmering ribbon that unwound. The next day at work, while I was still despondent, this message came to me: *Through life's trials and hardships we arise beautiful and free.* The vision of the butterfly returned to me.

That was when I began to learn how to internalize the process of the butterfly, which is all about understanding and letting go of our attachments. A caterpillar has a really comfortable life and grows attached to that comfort. But it's not truly free, it's not truly beautiful. Eventually, because it senses that there's something more—not by someone telling it, but by the deep intuitive force—it lets go of the comfort that keeps it grounded and spins a cocoon around itself. The cocoon comes from within the caterpillar, just as our letting go has to come from within.

The caterpillar encases itself within itself and is forced into this dark, small area, where it can't be distracted by anything. No longer can the sun and the rain enter its world. It is alone in the darkness, wrapped in what it has spun from inside, and shielded from any distractions.

So it is with us. True transformation occurs only when we can look at ourselves squarely and face our attachments and inner demons, free from the buzz of commercial distraction and false social realities. We have to retreat into our own cocoons and come face-to-face with who we are. We have to turn toward our own inner darkness. For only by abandoning its attachments and facing the darkness does the caterpillar's body begin to spread out and its light, beautiful wings begin to form.

Even then, the caterpillar must shed one last attachment— to the dark, cramped space it has gotten used to, a new form of comfort—and begin breaking through the barrier of self in which it has wrapped itself. It doesn't have a clue what lies beyond, but it responds to this higher calling anyway. This last struggle effects the final transformation. If a human helps the butterfly break through the cocoon, the butterfly will never fly. Only by finding the strength to break free of that last attachment can this delicate being, with a body so light and fragile that a breath could seemingly kill it, fly beautiful and free.

Similarly, only once we let go of all we know, including all our self-centered concerns, and break free of the cocoons we spin around ourselves to shut out the world can we become the truly beautiful beings we are meant to be.

❧

WHEN I ALMOST died in that mother of all storms, my fear of dying died, too. Letting go of that freed me, like the butterfly frees itself of its cocoon. I began to live day by day, moment by moment, breath by breath, and prayer by prayer. Before I knew it, I had reached the hundred-day mark I swore I would never see.

"I guess I made it," I giggled incredulously to myself as the press calls rolled in.

At a rally down in Stafford, I was presented with an award—as defender of the woods—by Leonard Peltier, a Lakota Sioux who had been wrongfully imprisoned for more than twenty-three years for a crime he never committed, and other members of the American Indian Movement (AIM). Their powerful drums resonated up the hillside. I was honored by the Native presence, because I so very much honor them and their way of life. My heart cries out at what we as white people have done to the Native people and how we continue to do it today, abdicating responsibility for the role we play in the continuing genocide of this dwindling population.

❧

THE VETERANS FOR PEACE also awarded me with their Wage Peace Recognition of Valor award for holding the line for one hundred days, saving the life of this ancient redwood.

Hundreds of people attended the ceremony. Watching them pop up on two main trails that lead up to Luna, I cried. It was amazing to me that all those people would hike up that steep hill to show support for the action I had taken. It takes a lot of love to make a person hike that hill! I felt honored and blessed.

I'd been regenerated.

I even became accustomed to living in a tree. We're such creatures of habit and routine that anything becomes natural after a while. Day by day, I adapted. I learned little ways to make life simpler. I organized the little space I had, using buckets with airtight lids and bags as storage. I figured out which way the wind usually blew and found ways to reinforce spots on the walls and the roof of the platform that would be most compromised. And little by little, things that seemed hard simply became my way of life.

In the beginning, my legs missed walking. I could feel them wanting to stretch out and stride. But eventually they wanted to climb instead. That's how I got my exercise. To keep up my strength, I also did push-ups and sit-ups as well as squats on Luna's limbs. But mostly I climbed.

At first, the only climbing I did was to fix my tarps. I started by wearing a rope and harness, but I never liked it. I just didn't feel comfortable with the gear; it made me feel disconnected. Even though it's a way of staying safe, it made me feel unsafe.

So after the first two weeks, I simply took my harness off and started free-climbing.

That's when I started learning to climb Luna with my fingers and my feet rather than by relying on my eyes. I could feel which branches would bend, because of the spring they had in them, and which were more likely to snap. I started noticing that the needles on the former were new and fresh, while those on the others were older, indicating fragility.

Gradually, as my frostbite healed, I started climbing around Luna to get to know her. I could do it only a little bit at a time because the weather was too cold and wet and my fingers and feet would go numb. But I kept at it, without harness, rope, or shoes, and learned how to disperse my weight between both hands and legs so that I never put too much weight on any one branch. As my comfort level grew, so did my desire to climb.

Since Luna splits in two about halfway up, I would cross over from one trunk to the other when I climbed, using the branches that grow between the two like a spiral staircase. I let Luna tell me where I could put my hand or my foot. When there weren't branches to hold onto, I looked for finger holds: little burrows, cracks, and crevices. Sometimes I'd use the knob of a branch broken off by the elements.

I explored the canopy, the upper half of Luna. There's a whole forest in her, and it's absolutely magical. Ferns, salmonberry, and huckleberry grow in Luna's pockets where duff has collected over the years. There are many different fungi and mosses and lichens; usnia hangs down like Spanish moss; scalloped, whitish gray lichen and teeny, tiny mushrooms shaped like satellite dishes nestle in her folds; green, furry moss, dark in the center and neon at its edges, coats her sides. Especially in the fog, Luna is a fairy tale waiting to happen.

Loggers like to argue that the tree is old, that she's just going to die anyway. But those parts of the tree that are dying are the very sites where different forms of life grow. And that circle of life is important for the forest ecosystem. Death is a part of life. Death feeds new life, which dies and turns to more birth. It is a magical, flowing cycle.

❦

AT FIRST, the idea of going to the very top of Luna was terrifying. The wind scared me the most. I always felt like I was off balance, that the wind was trying to rip me off the tree. So I'd wait until it wasn't blowing too hard before seeking out new pathways. At each precarious point, I'd have to overcome a new set of fears. My heart jumping out of my skin, I would gather my courage and trust that my bond with Luna would guide me in how, when, and where to climb.

Eventually I felt comfortable climbing behind the platform, which is not easy. To go above that, however, meant negotiating many branches that, having been exposed to such extreme weather, are thin and brittle. Still, I felt a constant calling to go to the top.

Maybe it was the feeling that drives people to hike mountains, that feeling of making it to the highest place possible, that inner urge to find our higher self. Some people think it's about an adrenaline rush or conquering, but I think it's a constant striving we have inside ourselves. I don't like the feeling of conquering. We conquer a mountain because we can climb it, then we conquer a mountain because we can blow it up for tunnels and highways and mineral extraction. We play God and destroy the natural balance.

So for me, climbing to the top was not conquering Luna. I just felt compelled to reach it. Once I learned how to disperse my weight among both hands and feet instead of placing it all on a single branch, I thought, "I can do this!"

So I tried. As I climbed in my bare feet—which I didn't wash so that the sap would help me stick to the limbs—I spoke to Luna. In return, she guided me to those branches that were safe and warned about those that were not. Without shoes, I could feel her underneath me and understand her messages.

I made it to her lightning-hardened pinnacle, the most magical spot I'd ever visited. Luna is the tallest tree on the top of the ridge. Perched above everything and peering down, I felt as if I was standing on nothing at all, even though this massive, solid tree rose underneath me. I held on with my legs and reached my hands into the heavens. My feet could feel the power of the Earth coming through Luna, while my hands felt the power of the sky. It was magical. I felt perfectly balanced. I was one with Creation.

No way could I allow Luna to be cut! Ever!

At the top of Luna.

LIFE IN LUNA wasn't always so transcendent. My birthday in February had been hard to handle. From the time we were fourteen or fifteen, my best friend, Valerie, and I had promised to always celebrate our birthdays (which were only two days apart) together. That way we would see each other at least once a year and always stay friends despite time or distance. This was the first year apart, and I missed her greatly.

But my birthday did help break the Redwood Curtain, as some people call it, the barrier preventing information getting out about what's happening to the remnants of our old-growth forests. Though we have less than 3 percent left—enough of a story in itself—it takes a human interest angle to get the media interested. In 1997, that human interest angle became me and my twenty-fourth birthday.

Enter the media world. Following a *Los Angeles Times* article in February, the tree-sit made it into the issues of both *Newsweek* and *People,* icons of the mainstream press, which I used to flip through while waiting for my doctor. That's when I realized what had happened.

"Oh, my God, I'm becoming a public spokesperson!" I thought. "I'd better climb down and get far away." I was not very happy about this turn of events.

It's not easy to become a public person when you've spent your whole life being private. But in trying to draw attention to Luna and the redwoods, my life was about to become an open book. Everything was fair game, from how I took a shower to how I went to the bathroom. We all use the bathroom, but we don't go around asking one another about it—unless you happen to be Julia Butterfly living in a redwood tree. Suddenly nothing was sacred, especially when the radio disc jockeys got involved.

John and Ken, shock jocks from Los Angeles, were the first to call while they were on the air. Shock jocks are paid to create tension and hullabaloo. Their whole objective is to make people go, "Oh! No way! Wow!" So when a photo of me taken with a fish-eye lens to get in all the surrounding greenery ran on the AP wire—a photo that also unfortunately distorted my head and hands and feet—they jumped all over me.

"You must be a hairy Neanderthal," they commented derisively. "You must stink so bad."

The abuse continued unabated. I did my best to come back with loving and respectful answers, but it was really draining. During one of the commercial breaks, I put my hand over the phone's mouthpiece and muttered something under my breath. A reporter who happened to be in the tree that day quoted me as saying, "These guys are assholes."

Having been raised in a religious home, I don't curse a lot. So though I was exasperated, I probably didn't use those exact words. But her article made me feel bad. So even though I didn't think she'd heard me correctly, I sent John and Ken a letter apologizing for my language. Naturally, John and Ken called me back.

"You want to go on the air in five minutes?"

"Okay," I agreed.

I didn't know if they were going to slam me all over again. I suspected they might, but this represented another opportunity to reach people.

"She called us assholes, but then she sent us a letter of apology for it," they announced when we got on the air. "Julia, you didn't have to do that. We get paid to make people call us assholes."

"Guys, I appreciate your good humor about this, but my message of love and respect is not something I take lightly," I answered. "I just want you to know that I'm sorry, because I don't treat people that way."

That seemed to blow them away a bit. They poked a little fun at me, but they didn't annihilate me as they had the first time.

During our third phone interview, we got along absolutely, wonderfully well. Then they couldn't call me anymore, because they don't get paid to get along with people.

But that didn't stop the rest of them.

One morning, at 5:30 A.M., the phone rang. I answered groggily, only to hear the sound of a chain saw.

"You better run! You better run!" screamed a couple of disc jockeys. "We're cutting you out!"

"Do you guys have no respect at all?" I stuttered, once my heart stopped pounding.

Thinking quickly on my feet became essential.

"You know, Julia, if you climbed up in a tree in my back-yard and refused to come down, I'd be really pissed at you, and

I'd do whatever I had to do to get you out," another radio disc jockey said during an interview.

"If you were one of the families that lived in Stafford, California, and your house disappeared under a wall of mud that began at a clear-cut and logging road, I think you'd have a whole new understanding of the word *pissed*."

That made too much sense. He got really quiet for a second and changed the subject.

The media and all that went with it had become a part of the Luna sit.

On April 5, as part of that first media wave, CNN scheduled a live debate between John Campbell, Pacific Lumber's beefy, Australian-born president, and me. *That* made me nervous. To go live on CNN and debate the president of the company I was protesting was a big deal and a huge responsibility. I got calls from all kinds of activists prompting me and making sure I knew what I was saying. I couldn't misspeak. I had to get everything right and state the case so that we came out looking good.

As it turned out, Campbell's inane comments made my job easy.

"Julia should really come down, because it's a large, old-growth tree, and a number of wildlife like to use that type of tree to nest in," he said.

"Your company marked Luna with a blue paint slash at the base, which means you wanted to cut her down," I fired back. "How much wildlife could Luna support if you turned her into somebody's deck?"

Afterward I got a lot of praise for my performance. But I didn't have to do much. The man clear-cut a path to his forehead and painted a bull's-eye on it.

Working in Luna.

THE SUDDEN MEDIA interest led to a whole new round of problems. From the start, the Luna action had been ragtag. A haphazard support team and a media office that moved from place to place made outreach difficult. Almond had done a good job for the little he had to work with, but as soon as the press got interested, the usual spokespeople for Earth First! jumped in. And they wanted to do it their way. They wanted their image portrayed in a specific way, and they weren't sure I was the best person to do that. In their minds, it was their movement. I was just some new upstart who couldn't possibly know what to say and how to act.

So Almond and I were supposed to do what everyone else was telling us to do and the way they wanted us to do it. With all the media attention after the hundredth-day rally, it got to be too much for Almond. He would get attacked repeatedly for not writing a press release the way they wanted. He was hanging in, but I could see that it was wearing him down. He would climb up in the tree, and I could see it in his eyes and hear it in his voice: his spirit was dying. I felt really bad for him.

Things began to get overwhelming. I decided to fast and pray. When you cleanse your body, you also cleanse yourself mentally and spiritually. So I fasted and prayed, because this tree-sit needed some divine intervention.

"Please send someone," I asked. "Almond needs a break, Almond needs to heal. We need someone with skills to help pull this all together."

Four days into the fast, I got a call from Robert Parker, a six-four, broad-shouldered, broad-hipped river guide and forest activist with a bit of a beer belly, a crooked bump on his nose where he'd broken it one too many times, and a lot of media savvy.

"Julia, I see that you guys need a central base point, a media center," he stated.

He had already prepared a list of his goals for the action and the time line he needed to accomplish them. A few days later he was on board.

Robert brought everything together in a cohesive way. He did outreach to the press. He helped establish a Web site. He set up an office where the press could call and actually find a human. He turned it into a well-oiled, international outreach machine.

The resulting media interest gave me an expanded purpose and a new life in Luna. When I climbed this ancient redwood tree, I never could have imagined that I was going to have my ear to a cell phone, my hand to a pager, and my other hand on a planner. But tree-sits have three purposes: to protect the tree and hopefully a few around it, to slow down the logging while the people who work within the legal system do their work, and to bring about broad-based public awareness. And that latter goal requires sophisticated technology.

In Luna, that has meant a radio phone powered by solar panels that are connected to two motorcycle batteries, an emergency cell phone, a hand-powered radio, a tape recorder, a digital camera, a video camera, walkie-talkies, and a pager that functions as my answering machine and controls my life. All this equipment was slowly brought in over time—one thing here, another there—all hiked up the grueling hill by a wonderful support team. Solar panels now hung from Luna's limbs, collecting precious sun and recharging batteries.

I didn't really like being at the mercy of electronic devices. My head spun all day and into the night. As long as my phone and pager were turned on, there were a hundred demands for my time. The fact that my pager number had been given out in a press release, and my phone number posted on a Web site, didn't help. Calls came in at every hour of the day and night. At some subconscious level I must have been feeling the need for a break, because right at the height of the media assault I accidentally dropped my pager. For months I had paid no penance to the gods of gravity, but now the pager fell one hundred eighty feet and shattered. Truth be told, I relished the few days of reprieve that it brought.

I believe that ultimately we're all going to have to wean ourselves from this dependency on technology. We can do it now by choice, or we can do it later when we no longer have a choice. All our growth is coming at the expense of the Earth, which gives us life. We have to live in a balanced, sustainable way. So I tried my best to keep balanced, but it was difficult. In order to get the message out, I had to keep my technology turned on. Up in Luna, I was living on the world's most amazing radio tower, which receives and transmits all the beautiful and powerful truths of our universe, and I had been blessed to be at the microphone on that tower.

Of course, towers do tend to attract energy from nature as well as humans. Luna was no exception. She had already been struck by lightning many, many years ago. With the other large trees around her cut down, and the metal solar panels at the top of her, Luna became a lightning strike waiting to happen.

One night I watched a storm, with its lightning and thunder, skip from one ridge to the next to the next, moving nearer and nearer.

"I've really got to think about coming down," I said to myself when it drew near. As much as I was willing to die for what I believed in, I didn't want to die for the sheer stupidity of sitting on top of a lightning rod in the middle of a lightning storm.

The storm got so close that the hair on my head and my arms was beginning to stand up.

"Okay," I said. "It's time to come down now."

I put on my harness and prepared to descend to where the ropes were so I could rappel down the rest of the way. Then the lightning hit, jumping and slamming all around me. "*Bam! Crack!*"

"I waited too long," I thought with dread. Then I felt the beaded medicine pouch that I'd been wearing since a woman named Sonya made and sent it to me and in which I keep objects that are sacred to me as well as sage and other herbs.

I don't know why or how, but the pouch started to heat up. It was like a message. So I responded.

"Great Spirit," I hollered into the storm, "protect us!"

And *wham!* lightning struck in the mud slide, right next to Luna. Everything shook. The sky turned a funky neon color from all the electrons running around, and my hair rose straight up in the air. I jumped and screamed. Maybe this was it!

But nothing else happened. The lightning suddenly jumped three ridges back and disappeared. I started laughing.

Once again, the power of prayer was working miracles in my life. Now, we needed a miracle for Luna.

NINE

CROSS-FIRE

Springtime in Luna. I felt truly blessed. I awoke to nature's morning breath, sweet as honey, and watched the sun rise, a gorgeous red and orange and peach and then gold, shooting across the fog in the valley below, catching it on fire. There were so many scars on the earth around me, but the fog covered them over like a blanket. I felt like I was on the ocean, with islands of mountains and ridges rising in the distance.

Up in the rain forest canopy, days of sunshine are rare. At the first sign of warmth, the living creatures around me surfaced. Insects burst forth out of all the cracks and crevices, dropping on me from everywhere. Centipedes in the duff of the upper caves, reddish colored, with two little prongs coming out of their heads and a bunch of legs on either side, squiggled around. Little winged bugs of all different shapes and sizes and colors, reds and browns and blacks. One bug had purple iridescent wings, so thin they didn't look real. Another had fairy wings of iridescent green, the color of green leaf lettuce, brilliant, shimmering, almost fluorescent.

Bugs!

Before I came into Luna, I didn't care what kind of bug it was—an ant or a spider or a mosquito or a tick—I would kill it, kill it, kill it. I always loved being outside, but the one thing that bugged me was bugs.

Spiders gave me the true heebie-jeebies. My brother had been bitten by a brown recluse, and the poison ate a hole in his thigh. He had to get eighteen shots to avoid losing his leg. Another friend of mine back in Arkansas almost died when a spider bite on his neck swelled up so badly that it cut off his oxygen. He was in critical condition in intensive care for two weeks. So spiders were hard for me.

In fact, getting used to living with any kind of bugs has been a bit of a stretch. One evening as I was talking on the phone, a daddy longlegs fell off a branch and landed on my cheek. That flipped me out. My initial reaction was to smash it, but I knew I couldn't do that anymore. It was going against what I'd been learning here in Luna, that all living things have the right to be alive. So I plucked the daddy longlegs off my cheek, moved it to a safe place—where I wouldn't accidentally squash it—and wished it well. I began to do that all the time. I'd let insects crawl on me or gently guide them in another direction. They just didn't bother me anymore.

All sorts of living creatures dropped into my life on that platform. Northern flying squirrels seemed hell-bent on making my nights as interesting as possible. Just as I would fall asleep: *crash, boom, bang.* They have the funniest little chatter, too, as if they're talking on an old forty-five record switched to the wrong speed.

Because of my tiny space, many things were hanging and dangling and stacked, and the squirrels made it their highest priority to knock over everything they could, hitting and clanging every accessible piece of metal along the way. Nightly, they would wake me up. I'd stay snuggled in my little cocoon, hoping they would settle down, but they rarely did. Finally, I would get up and tell them to go away.

"Okay, kids, it's time to get to bed. Or at least let mom go to sleep."

They would take off for a while. As soon as I would nestle down and start drifting off to sleep, however, here they would come all over again.

With springtime they seemed to get an injection of adrenaline. If I could have gotten nocturnal with them it wouldn't have been so bad, but I didn't have that ability. My sleeping time was for them the time when all hell should break loose.

I couldn't even get back at them, since this was really supposed to be their home, not mine. I never did find their nest. I did, however, discover caches of edible tidbits in various nooks and crevices and caves.

They ate anything. I worked really hard to not tame them so they wouldn't get hooked on my food, but they would find my little droppings anyway. I used to sit real still and watch them. My carpet was just like a blanket, and they would push the wrinkles aside and find a couple of grains of couscous and other goodies.

They were pretty intent on getting food. Northern flying squirrels seemed slightly barbaric to me. They don't understand sharing. The pair I lived with in the tree used to nag each other all the time. When the female would find a scrap of food, the male, always overbearing and rude, would want it. So he would make these loud noises at her and then lunge. She'd fly up the tree, hang from a branch, and peer at him upside down.

When I first arrived at Luna, the logging below coupled with the helicopters above had driven away the animals, so I had no competition for my provisions. Once the logging operation was over, however, the animals popped up, returning from their necessary exile, and my food was fair game. I tried storing my supplies in buckets, but they ate through the buckets. I finally resorted to wrapping everything in plastic bags to cut down the smell and then putting the bags in the lidded buckets, which worked pretty well.

Once after my northern squirrel cohabitants had chewed through the bucket and nibbled bites of four different carrots, I gave in and set one out for them. Maybe they'd leave the rest alone, I reasoned. The female appeared and started gnawing on it until, as usual, the male chased her off. Then he proceeded to try to carry off the whole carrot, which was as long as he was. Stumbling, he tried to make his way through the assortment of little jars I had set out. Time after time, he would hit them and get stuck.

I sat there and watched to see if he was going to take the time to nibble off a section he could manage, or if he was just going to keep on being a greedy, overconsumptive jerk. I finally had to take some direct action. So I took away the carrot, broke it in half, and gave half to him and the other half to his partner. Finally able to manage his load, he scampered up the tree with his bounty.

I guess I was imposing my own ethical standards on those squirrels, because I felt bad for the female. He was just so obstinate and mean, always picking on her. I guess that's the way it's supposed to be. It's nature. I imagine she wouldn't be with him if that really bothered her.

⌒〰⁓

THE NORTHERN FLYING squirrels provided wonderful entertainment, but the mice were funny, too. They were just much harder to keep up with because they're so tiny. Most of the time, all I could see was a little tail zipping around.

One of my most amazing visitors were hummingbirds, because people say they're not supposed to fly that high. But they do. I would sit on the platform minding my own business, and suddenly I'd hear *nyearw, nyearw,* and a hummingbird would fly right in front of the opening and hover straight in front of my eyes. "My God," I thought to myself, "how can they flap their wings that fast and not get tired?"

Springtime in Luna brought all sorts of birds. There was a brown, delicate one, either a sparrow or a wren, and another

© JULIA BUTTERFLY HILL

"Hey, that's my lunch you're eating!"

darker one with a pointy plume. There were finches—the male, a bursting ball of sunshine yellow, the female with a soft, light gray bottom and a yellow top. There were chickadees, which look like balls of brown and black and white cotton with a beak stuck on them. They had the most beautiful songs, melodious and sweet.

There were also lots of turkey vultures and ravens, always proof that an area has been destroyed. These scavengers don't exist in an intact forest system, because they can't find their food. But when logging opens the forest up, they can catch their prey.

The vultures would swoop around and around without flapping their wings. They were so beautiful from afar, these birds of prey, but they served as a reminder and a warning to me that things that may seem beautiful could be waiting to prey on me. The materialistic, mainstream world looked freeing and desirable, but it, too, was like a bird of prey. To the media, I was a commodity—their quarry. They were looking to feed off me. The press covered me because it had decided I was sellable. I had done my best to turn that around and use the media as a tool to get the message out, but the swoop of the vultures was a reminder to me to stay on the alert.

Unfortunately, I was never blessed with the presence of spotted owls or marbled murrelets, although there were many horned owls hooting in the shadowy night. They need an old-growth forest that hasn't been denuded. Reducing the forest to a tree here and there and destroying most of the old growth does not leave the proper nesting habitat for either bird.

Marbled murrelets are extremely susceptible to predators. When healthy and thriving old-growth forests lined the coast, they could fly a very short way between the ocean and the safety and cover of a canopied forest. But now, because of our refusal to understand the interconnectedness of everything on the planet, we have fragmented their habitat. Leaving isolated groves or single old-growth trees isn't enough, so the marbled murrelets cannot nest in what is left around Luna.

The swiftly falling numbers of northern spotted owls and marbled murrelets show clearly the cumulative impact of destructive human practices. A clear-cut here and there adds up to much more than the sum of its parts. That's the importance of cumulative impact. We have to realize that the fragmentation of our planet is affecting the quality of life for everyone and everything. Political and private property boundaries are destroying nature by dividing it up into separate pieces that don't make any sense. We must find ways to fit our lives into the natural ecosystems and watersheds.

The coho salmon are also being decimated by our refusal to respect the cumulative impact of our conduct. The timber industry points its finger at commercial fisherman, commercial fishermen point their fingers at the timber industry, and both of them point their fingers at the seals in the ocean when all else fails.

But it's all about cumulative impact. Every action we take on this planet affects so much more than our personal lives. If we recognized and accepted responsibility for the compounding impact of our actions, the streams and rivers where the fish spawn would not be overloaded with sedimentation. The people looking to the salmon for food would take enough for subsistence and would not take for profit. If we accepted our responsibility for the cumulative impact we have on the environment, there would be plenty of fish for all life, including the seals.

⚬⁓

LIVING IN A TREE this size, I felt balanced right at the center of Creation, the way we are *all* balanced at the center of Creation. I learned firsthand how everything that we can—or cannot—see is interconnected like strands in this web of life, from the microorganisms in the soil helping feed nutrients to Luna, to the stars that are billions of light-years away, and everything in between. Living the way I live, I've learned how each thread reaches back to us.

One day, as I was climbing around Luna and the fog slipped back to the coast, sunlight hit a spiderweb still glistening with drops of moisture. It shot these beautiful spectral colors in every direction. The diversity of life is like those strands of the spider's web: the strands weaving together make the web strong and balanced and, even more amazing, make it beautiful as well.

IN LUNA, time flowed like the Eel River, the silver ribbon that winds through the valley below. Some days it meandered slowly, and on others it rushed headlong into the ocean. But it always flowed instead of being segmented by a calendar. Before I knew it, six months had gone by. That's when it hit me.

Just a few months earlier, I had been told repeatedly that I couldn't be of use. Two months into the Luna sit, I didn't even think I could make it a hundred days. But, somehow, half a year had passed, half a year without walking on the ground, without doing all those things I used to take for granted. It no longer even felt like much of a sacrifice.

"Is it worth saving this one tree?" everyone kept asking. "Isn't the hillside completely raped?"

I never understood why they all focused on the negative. Even if Luna had been the only tree left, yes, she would have been worth it to me.

BY THE SPRINGTIME, Pacific Lumber had pretty much given up on security guards. One day, I saw all these people pouring in from about half a dozen different trails leading up to the tree. They gathered in a beautiful, colorful circle in the clearing two hundred feet behind the tree. Mickey Hart, formerly of the Grateful Dead, had hiked up the hill with his drum and his core group of drummers, and he initiated the main beat. Then everyone else mixed in with bodies and hands and instruments and voices. There must have been a couple of hundred people playing on drums and flutes and pots and pans and shakers and everything else imaginable. We celebrated the sustainable future for everyone that we were working toward, and we celebrated our love for life. It was really windy and quite cold that day, but looking down on this wondrous scene, I felt wrapped in a warm blanket of love and commitment.

To have Mickey Hart visit Luna, and to feel his energy and support, was a big deal for me because I've always been a Grateful Dead fan (though not a "Dead Head"). To say I was deeply honored by him and everyone else is an understatement!

I climbed to the top of Luna for the party. When they started drumming I started dancing, feeling the beat, letting my arms be branches, bending and flowing with the music. In the tree I can't dance normally; I have to dance either from the waist up or the waist down rather than both at the same time.

There were two spots on the tree where I liked to dance. On the very top I danced from the waist up while I held on with my legs. A little below that there were two large branches. I'd grab one with my arms and groove from the waist down.

So I fluctuated back and forth between the two, and when I'd get too tired I'd go sit out on a branch and watch all the other dancers and drummers. Everybody was feeling so good, so full of love, so full of joy, and we danced that beat into the Earth, and we beat that rhythm into the heaven. Luna and the forest resonated—such an immense high.

By dancing, drumming, and even hiking up the hill with good feelings in their hearts, those people were sharing their love with the hillside. Nature needs that. Nature gives love to us every day, but we've forgotten how to listen, we've forgotten how to communicate, we've forgotten how to give back.

The tree-sit was definitely attracting a different, more visible group of supporters. Shortly after the spring celebration, Woody Harrelson climbed the tree for a personal visit. Media world now started to fuse with celebrity world. He arrived on a hectic day, so only later did we have a chance to really start talking—about his activism, about how he got involved, about his family. We stayed up half the night.

He ended up staying the night. I didn't have that many blankets, and it was extremely cold and windy, but I did my best to make him comfortable by giving him half my covers. Neither of us stayed very warm, but I was more used to the cold than he was. He had difficulty with the sparse situation, and when the phone woke us at around six the next morning, he groaned profusely. He's obviously not a morning person. Later I would find out that even his closest friends, family, and assistants would rather walk on hot coals that have to deal with waking him!

"This isn't a five-star hotel," I said when he jumped on me about the call. "You're in a tree-sit. This is a campaign. We do things like deal with the phone at six in the morning. I don't have a personal assistant to handle things as you do!"

"I'm sorry. I guess I'm a little spoiled," he said with a grin once he'd gotten over the sleepy grumpies.

Laughingly, I agreed. I truly appreciated the fact that he was real, though, even if his reality was very different from mine. His activism, encouragement, and support meant a lot, not only to me, but also to the forest protection movement.

"I don't see how anyone could hang out among these magnificent trees, some of the oldest and biggest living creatures on the planet, and not be alarmed by the prospect of them being logged," he subsequently told a local newspaper reporter.

THE IRONY WAS that while living in a tree, I, too, had stumbled into becoming a public figure. I didn't like that much. But if I *was* to be a spokesperson, I wanted to make sure I was an informed one.

On my very first day in Luna, while talking to loggers, I had quickly figured out that I needed more than heart and instinct to make my case. I didn't really know the difference between late successional forest and residual old growth forest. I didn't know that Luna was the only ancient tree in the area, probably spared because the mills at the time the area had originally been cut couldn't handle a tree with such curves and twists. The only thing I knew was that the forest was being turned into clear-cuts and mud slides, and that was wrong.

I set about educating myself on the issues. I listened to environmental radio shows. I requested information from EPIC (Environmental Protection Information Center) and the activist collective Trees Foundation. With a desire to learn as much as I could, I talked to experts about slopes, instability, watershed analysis, and timber harvest plans. I read reams that people downloaded off the Internet and brought to me about sustainable logging in the forest—and about Charles Hurwitz's financial dealings as well as Pacific Lumber's decision to double and in some cases triple the rate of cutting in order to maximize profits. I studied up on local northern California history and a hundred or more years of timber history. In short, I gave myself a crash-course graduate education in environmental issues. I was even presented with an honorary doctorate in the humanities from the California-based New College!

"What's up, Doc?" my friend and longtime ground support, Spruce, asked the moment he found out about the honor. The notion of me being a hundred and eighty feet up in a tree with a doctorate was one he couldn't resist.

BY MANY GAUGES, the tree-sit had been a tremendous success. It had become known across the country, even around the world, informing and inspiring many. Incredible press coverage, both domestic and international, had helped to legitimize direct action in mainstream society, which represented a huge shift away from the earlier negative stereotyping of *environmentalists*. Already, these factors alone had taken the tree-sit far beyond my wildest imaginings and expectations.

But many people continued to ask, "What has it really accomplished?" A thousand-year-old tree, condemned to death, had been granted a reprieve for six months, while the devastating clear-cuts and the destruction of old-growth forests continued all around me.

The largest local paper, the *Times-Standard,* published in Eureka, California, ran an editorial entitled "Time's Right to Put an End to Protest." They said:

Six months in a tree is an astonishing physical accomplishment, which hardly needs to be prolonged. There is a curve of diminishing effect in attention-seeking—the most arresting novelty eventually becomes stale news. A year in a tree is not necessarily a bigger story than six months in a tree, and two or three years ceases to be news at all.

That's what the local press thought of this tree-sit—it was a story, a novelty, an attention-getting device. It was narrow-minded and myopic. They called it "an open-ended endurance test," and they said that Luna was not "an especially notable tree" because it did not set any "size records."

How could they have so totally missed the point? I did not go up to live in Luna because she sets any size records. I did not stay to set any endurance records. To me, the tree-sit was not about records of any kind. It was not just a "novelty." It wasn't even about me. If it had been about any of those things, I would have gone down long before this point. I had put my life in a critical position to try to show people what was really at stake. These old-growth forests, once destroyed, would never return. This was not to gain a spotlight for me but to shine a beacon on something that was about to vanish forever. It was about the forests.

That week when I spoke to Geraldine, host on public radio KHSU, who has included me on her weekly radio show almost since the beginning of this tree-sit, I told her that I thought the *Times-Standard* was wrong. The time was not yet right for me to come down.

"Pacific Lumber has not yet promised that Luna will be allowed to stand, and I will stay until they do," I told her. "I do not want to come down to a world where clear-cuts are allowed, where herbicides are dumped on forests and on people, where logging continues on unstable slopes, where old-growth forests are destroyed. Why should I want to come down to a world like that? I still have my work cut out for me up here in Luna. I need to get people rallied to stop those things."

Nobody wanted me to come down more than Pacific Lumber/Maxxam Corporation. They couldn't continue liquidating these trees away from the public eye with me sitting up in a tree. By this time, however, they had given up trying to scare me or intimidate me. Instead, they figured they could outwait me.

"She can stay up there as long as she wants," Mary Bullwinkel, Pacific Lumber's spokesperson, and John Campbell, the president, started telling the press. "We are going to let her stay until she's tired and comes down."

Like the editor of the *Times-Standard,* they just didn't get it. They didn't realize that they would not be able to outwait me. I would not come down without some progress being made toward saving the forests.

My father, on the other hand, completely understood. "If John Campbell thinks he can outwait my daughter, he doesn't know my daughter," he said at a press conference in July 1998. "Every time her mother and I tried to control her or outwait her on something, we always lost."

Still, I needed to do more than outwait them. So having already taken my message to the media, I moved on to the government. I began speaking out against the Headwaters Forest Agreement, which was a proposed plan to save approximately 3,500 acres of intact ancient redwood forests at the expense of species on the verge of extinction and thousands of acres of old-growth forest. On the surface, to people who didn't really know what was at stake, it looked like an eco-friendly enlightened proposal. I knew as much about the issues as anyone at this point, and, having proven my commitment, I wanted to be of service beyond saving Luna alone.

Robert Parker, in his capacity as media and ground support coordinator, drove down to Sacramento with a cell phone and ESP (Darryl Cherney's Environmentally Sound Promotions) and set up appointments for us with eight key members of the state assembly and senate. Robert would meet with them in person, then he would dial me up and push the speakerphone attachment on the cell phone.

"The government has a responsibility not to turn its back on the local residents who will be affected by this agreement," I told them. "Seven families below me in the town of Stafford no longer have homes because of logging practices that will be allowed to continue unchecked if the plan goes through. How many more families will be made orphans of this so-called agreement?"

Sometimes the lawmakers gave us two minutes, and sometimes they gave us twenty, depending on how many others had come before us and how irritated they were. One of them, state senator Mike Thompson, who represents the district in which Luna stands, called me back from the senate floor. I could hear them doing a roll call vote in the background. The issue was the state's appropriation for the Headwaters Forest Agreement.

"It's not a wise use of taxpayers' money," I told the legislators. "It's not true protection for the environment, for a sustainable economy, or for people's lives. It's not a good deal no matter how you look at it."

Why did I oppose money that would supposedly be spent to preserve old-growth forests?

The Headwaters Forest was "discovered" and named in 1987 by forest activists who found out that Pacific Lumber, which had been taken over by Maxxam Corporation two years earlier, was clear-cutting a huge old-growth forest, the largest stand of ancient redwoods not owned by the government. Activists from Earth First!, EPIC, Trees Foundation, Bay Area Coalition for Headwaters, and all sorts of other groups tried to stop the logging through enforcement of the Endangered Species Act, since Headwaters was prime habitat for many species, including the endangered northern spotted owl and the marbled murrelet.

In 1996, Dianne Feinstein, the U.S. senator from California, announced that an agreement had been reached between government representatives and Maxxam. The federal government and the state of California would pay $380 million to buy Headwaters and some other lands that Pacific Lumber had already logged. Later, two other tracts were added, and the price was raised to $480 million.

The problem was that Charles Hurwitz already owed the government over three times that amount. Maxxam had been accused of looting from a savings and loan. The bailout, some say, cost the federal government about $1.6 billion dollars. A swarm of litigations followed the collapse of Hurwitz's United Savings of Texas.

Now the government was offering to pay him even more money, which he could use to buy up more timberland and continue his destructive practices. He had held the Headwaters Forest as hostage until the government agreed to pay an exorbitant ransom.

The proposed agreement called for Pacific Lumber to prepare a Sustained Yield Plan saying how much it would cut over the next hundred years. According to its Sustained Yield Plan, Pacific Lumber would do most of its cutting over the first twenty years, then reduce that rate drastically over the next eighty to achieve a "sustained yield." So instead of cutting a reasonable amount over the entire hundred years, they would basically annihilate everything into near oblivion in twenty years, then lay off workers. That's not my concept of sustainable.

As part of the deal, the company also had to prepare a Habitat Conservation Plan, which described how it would manage the rest of its holdings for the next fifty years. As it turned out, their Habitat Conservation Plan was simply a loophole to get around the Endangered Species Act. In essence, it said, "We know species are in danger, but if you protect them in some areas, we'll let you destroy them over there on 200,000 acres." According to this plan, the Endangered Species Act would basically be null and void on all of Pacific Lumber's land. Pacific Lumber would be allowed to destroy endangered species by eliminating their habitat. They call that an "incidental take."

That didn't sound very incidental to me, especially since it was built right into the plan. And what *incidental* means to someone whose company has been cited more than three hundred times for violating the California State Forest Practices Act is probably very different from what *incidental* means to the average person.

In my opinion, this was no way to preserve species on the verge of extinction.

When a species is in critical condition, you should be restoring, not destroying. Yet this proposed agreement provided nothing for restoration. They said the Habitat Conservation Plan was based on the "best available science," but how can you call it science when you compromise a species that is in danger of extinction? No amount of science can make a bad concept a good one. That's why, in my mind, Pacific Lumber's HCP (Habitat Conservation Plan) stood for *Horrible Concept, Period.*

The Headwaters Agreement would allow thousands and thousands of acres of redwood forests, as well as beautiful Douglas fir forests, to be cut into oblivion. Once Pacific Lumber was done clear-cutting, it would be able to dump herbicides and diesel fuel, the carrying agent for the herbicides.

There were a few good things written into the agreement, like large buffer zones along the streams. But how would this be enforced? By an assessment team made up of five members—three of them appointed by Pacific Lumber. Pacific Lumber was supposed to police Pacific Lumber. The robber was to become the cop. It asked us to believe that they had become suddenly "rehabilitated."

To me, that didn't make a whole lot of sense. Pacific Lumber has had a difficult time living up to the law, let alone enforcing it. It was cited for hundreds of violations of the California Forest Practices Act in three years—so many that its license to cut had even been suspended! (I found it ironic that Pacific Lumber tried to portray me as this horrible law-breaker, when I didn't even have a criminal record, while their company had a huge list of violations.)

"There *were* many violations," Mary Bullwinkel, Pacific Lumber's spokesperson, admitted when I debated her on *Time*

Online. "We have hired new people to prevent the same thing from happening in the future."

But it still happened over and over and over, and still the company wasn't punished.

Let's say you lived in a city, and, over a span of three years, you neglected to put enough change in the parking meter on three hundred occasions. As soon as the authorities caught up with you, you would have your license suspended and your car impounded, and you would probably wind up paying the resulting fines for the rest of your life. For mere parking violations. Meanwhile, Pacific Lumber's so-called minor violations were destroying critical habitats, threatening endangered species, causing massive mud slides that wiped out people's homes, and destroying air and water quality, which every species on this planet needs in order to survive. And what happened to them? In one instance, they were fined $13,000, which they could pay without blinking. They could make several times that amount every time they broke the law. It was simply more profitable for them to break the law than to adhere to it.

❧

THE HEADWATERS AGREEMENT, with its bogus Habitat Conservation Plan and Sustained Yield Plan, was so bad you had to laugh, because otherwise you'd cry. I tried to tell all this to the senators and representatives, or at least as much as I could in the few minutes they gave me.

"We'll take the Headwaters Forest, thank you," I said. "But the government and Pacific Lumber can keep their bad deal. We'll take the forest that should have been protected long ago, but we're not going to take the deal."

I suggested instead that they set aside the money for forest preservation until they came up with a better plan. In fact, some foresters and activists had prepared a Headwaters Forest Stewardship Plan, which would have saved Headwaters without leading to the rape of all the rest of their land.

Did the government listen? Barely. The public got its chance to make a token response, but it wasn't lasting and it wasn't real, because public concerns—about indigenous rights, about losing homes to mud slides emanating from clear-cuts, about herbicides leaching into water systems—should have been included from the very beginning but never were.

"You act like an expert on so many things, but as far as I can see you are simply a lady living in a tree," challenged a caller during one of my radio interviews on the topic.

"Yes, I am definitely a lady living in a tree," I agreed. "But that is the key element. I have been here seven months now. It has given me a level of expertise because I have been able to witness things from a vantage point that no one else has."

And from my vantage point in Luna, it was absolutely clear that the Headwaters deal was a bad deal for the rest of the land that Pacific Lumber claimed to own, including Luna. Headwaters would be saved, but Luna and the rest were on the slaughtering block.

DESPITE THE INCREASINGLY public visibility and voice that my presence in Luna had granted me, Pacific Lumber never responded. I was bringing such a big spotlight into this area, I felt sure they would do whatever they could to get me down. But they didn't. John Campbell, Pacific Lumber's president, still wouldn't return my phone calls, which I'd started placing regularly in March. I felt sure that something big had to give.

Robert and I thought we had enough support to pressure Pacific Lumber into making some kind of move to protect Luna and the surrounding area. So, with the help of lawyers, we drafted a statement that called for the protection of all the remaining old growth in the area as well as a small buffer around Luna. In addition, we specified that all logging in the area would have to be done under the guidelines of the Institute for Sustainable Forestry. We presented it as a resolution supported by noted performing artists and political and community leaders—including California state senator Tom Hayden, environmental leader David Brower, and musicians Willie Nelson, Lacy J. Dalton, and Merle Haggard—and faxed it over to the main office. Then Robert and I, in a three-way phone conversation with Mary Bullwinkel, said that we were hoping the company would take a look at it and get back to us.

Still no response.

We called again.

"We just want you to know that we'd love to hear back from you, but if we don't hear anything, we're going public with this at a press conference."

Nothing.

So in July 1998 we held a press conference to publicize the terms for resolution we'd offered as well as the support we'd lined up. The move blew up in our faces. The local newspaper quoted Campbell as saying that I was making nonnegotiable demands, and that was basically like negotiating with a terrorist.

On the flip side, some people from within the movement jumped all over me for negotiating with the devil. They got many people to sign a letter stating that Earth First! doesn't negotiate and that I wasn't speaking for them, which of course was exactly what I had been saying all along. As a friend of mine told me, "Don't worry, Julia. For every action there's an equal and opposite criticism."

Strangely enough, that's what it took for Campbell to finally start answering my calls. I had phoned as usual, and his secretary took a message as she always did. But this time, Campbell called me back. I was floored—as floored as you can be when you live in a tree.

"Did you look over what we sent you?" I managed to ask.

"Yes, and there's no way I'm going to go for that. You're demanding too much."

"Well, that's just my initial vision of terms for a resolution," I said. "I have to tell you that I'm a little upset that you went on public record saying that I was making nonnegotiable demands, when I never said anything of the sort. All I said was, 'Here are terms for resolution. If you agree to them, I'll come down.'"

"Whatever you call them, I can't agree to your terms," he countered.

"If you don't like that resolution, is there any way at all that you can see saving Luna?" I asked.

"No. I will not negotiate with a lawbreaker!" he exclaimed. "If I negotiate with you, then people will be climbing up in all the trees."

"Mr. Campbell, the reason people are sitting in your trees is that you're cutting them down," I argued. "People are going to continue sitting in your trees as long as you're cutting them down. I'm not the first person to sit in your trees, and I won't be the last. The difference here is that this action has gained a spotlight that no tree-sit has ever had before. Because of that spotlight, you and I are both aware that I have a leverage that you've never had to deal with. This isn't like setting a precedent, this is like making history."

Campbell had had enough.

"Well, I'm open to continue talking to you, but I'm not interested in protecting Luna or the surrounding area at this time," he said.

But at least the dialogue had started. Campbell began taking my phone calls.

IF I HAD SEEN what the Luna tree-sit had in store when I first got involved, I would have run screaming in the opposite direction. But, like a stepping-stone or the rung on a ladder, each hardship taught me what I needed to learn so that I could reach the next step and the next and the next. Somehow, they seemed to come in just the right way, giving me the knowledge to handle, or transform, whatever came after. The universe may not always send us what we want, but it always sends us what we need, and sometimes a little bit more to make us stronger.

"Perspectives"

Pacific Lumber Mill
Mars the night sky
with its pitiful attempt at
pretense
Tail lights of passing cars
on highway 101
appear to be an army
of working ants
disappearing around
the curve and ridge
seemingly swallowed
in the Redwood Empire

8/30/1998

TEN

LIFE AND DEATH

I was pregnant, at least according to a rumor that resurfaced upon my nine-month anniversary. I even got a call from a group of midwives offering their services. I figured that any day I'd be giving birth to baby redwood trees!

Despite such silliness and our recent political fallout, the word about the action I had taken in Luna was gaining more legitimate recognition in the mainstream press. In August 1998, *Good Housekeeping* nominated me as the one of the most admired women in America. Me—the same person who had been called crazy, wacko, extremist, far left, terrorist, idiot. That proved to me that social consciousness regarding direct action and civil disobedience was beginning to shift. Suddenly, people were beginning to understand that I was not this insane radical. I, as well as all those like me who have opted to put their lives on the line for their beliefs, simply felt compelled to take a stand because of the blatant destruction of the Earth and ourselves.

And there was still so much to stand up against. While all the attention was focused on Headwaters, Pacific Lumber had stepped up its cutting in other areas—places like the Mattole, with its seven thousand acres of old-growth Douglas fir. The idea was to get in there, get the activists who had stepped up their preservation efforts out, and get the trees down.

We hear a name for a place—a name like *Mattole*—but we don't fully realize that these names are made up of residents and history, and these people have blood running through their veins. They have hearts that feel and minds that think, and their lives are being truly devastated.

The Mattole was one of the many areas scheduled to be sacrificed in the Headwaters deal. It has exceptionally high seismic activity they call the Triple Junction because three earthquake faults intersect there. It also has the highest rainfall in California, which makes it exceptionally prone to landslides. So you would think that the California Department of Forestry, which I call the California Department of Logging (because it seems to do much more for the timber industry than for the forests), would understand that logging there wasn't a sound idea. But the department doesn't seem to care. It has a big rubber stamp that says *Approved,* and the stamp gets put on just about every Timber Harvest Plan that Pacific Lumber submits. It approved the Timber Harvest Plan for the area that resulted in the giant Stafford slide. Then it approved the Timber Harvest Plan for Luna, even though we sit right next to that slide. And it approved the Timber Harvest Plans in the Mattole.

The Mattole River is home for the coho salmon, which is now listed as a threatened species. When the National Marine

Fisheries Service found coho in the Mattole, it took a really strong stand. Usually, the fisheries service releases statements about how logging will result in the loss of fisheries, which the California Department of Forestry blatantly ignores. This time, the National Marine Fisheries Service said, "Wait a minute. This is horrifyingly wrong. This could constitute a take." Under the Endangered Species Act, a *take* means a taking of habitat that could cause the death of a species, meaning that species would go extinct.

Pacific Lumber agreed to wait for further scientific review. But before long, they were cutting again.

ON AUGUST 13, 1997, Michael Evenson, a rancher from the Mattole, filed suit. He knew that he had enough information to stop the logging if he could get an open-minded judge, one who wouldn't cave to timber pressure. But while the Mattole's residents were waiting for the court to move at its customary snail's pace, their backyards were being destroyed. So, together with forest activists, they opted to take direct action. They gathered at the gate to stop the loggers from entering the woods.

The demonstration bought them a little time, but the sheriffs came and the loggers eventually got through. As it turns out, some of Pacific Lumber's newest employees were not out there to log; they had been sent out specifically to get the activists. In a change of policy, Pacific Lumber had decided to hire and train men to deal with activists using "pain compliance tactics."

These men behaved decently with the residents at the gate, but once they got under the cover of trees, they behaved like thugs and declared open season on the protesters.

They chased Nature Boy up a tree, then went up after him. Though he was high in the air, they twisted his wrists and arms to get him to let go of the branch he was hanging onto. Then they duct-taped his hands behind his back, tied a single rope around his waist, and lowered him out of the tree without a safety attached to the rope. The single rope hurt him because it was literally cutting him in half at the waist. With his hands duct-taped, Nature Boy could not protect his head against the branches as they lowered him down. He would not have been able to protect himself should the rope have failed.

They chased a young college student named Orange up a tree and proceeded to cut the tree out from under her. She had to jump from a height of thirty feet when they started sawing through the trunk.

They did that to Sawyer, one of the main activists responsible for building Luna's platform, too. He and another activist were fairly low down in an old Douglas fir, the next one in line to be cut down. The loggers started cutting the tree right between the two of them. A single slip of the chain saw would have taken Sawyer's leg off. Another logger took his chain saw up to Sawyer's neck. It wasn't turned on, but it was still extremely sharp, and it took a chunk out of the top part of Sawyer's sweater, right by his neck.

It was nuts, absolutely insane! In today's judicial system, people sometimes get off a murder charge by pleading insanity. But how can violence be anything but insane? Ultimately, there are ways to work through our problems that don't include violence. Violence on any level is not okay. We need to try to understand one another, not attack one another. Unfortunately, many people still don't seem to understand this.

"What would you do with Julia Butterfly?" a man was asked in a televised man-on-the-street-opinion interview.

"Cut the tree down. If she dies, that's life," he answered.

That's not life. That's death.

❧

IN HISTORY, it seems most people are honored only when they are dead. We have to change that perspective. We need to understand how valuable things are while they are still alive.

Out in the Mattole, Pacific Lumber employees were beginning to forget the importance of life. In August 1998, I told Geraldine from KHSU radio:

What is happening out in the woods right now, people's lives are in serious, serious danger. When you are using tools like chain saws and when you are cutting down trees, one small move and a person dies. We are talking about life or death, and I think it is time that we stand together in love to celebrate life, instead of being angry and potentially bringing about the death that we are seeing in the forests. God forbid if it takes someone's life.

It was as if I'd had a premonition. Just over a month later, on Thursday, September 17, my phone rang at 2:45 P.M.

"Gypsy's been killed," sobbed Felony, my normally joyous friend and fellow activist. I'd never met Gypsy. But I knew he was a forest activist.

"How do you know?"

"Because Farmer came out of the forest screaming and crying, saying, 'They split his head open! They split his head open!' "

Felony didn't have any absolute proof, but she was freaking out. I made a couple of phone calls to others and made sure they were aware of the potential disaster.

❦

DAVID "GYPSY" CHAIN, along with some other activists, had been out trying to stop logging operations that were being carried out in an area too close to endangered species within a Timber Harvest Plan. Gypsy and others had called the California Department of Forestry and asked it to show up. But the department has the tendency to come after the forest is already gone, find Pacific Lumber in violation, and issue a slap on the wrist. Meanwhile, forest has been destroyed. That's why activists go in and try to slow down the logging. They want some trees to be left standing by the time the regulatory agencies show up or the courts give a ruling.

It doesn't take long to fell an entire forest. With chain saws that are six or seven feet long, loggers can now mow it down in a couple of days. So activists go out in the forest and literally put their bodies on the line. They do what's called cat-and-mousing. They run up to the loggers and engage them in dialogue, then run off if the loggers try to chase them. Loggers are not supposed to fell trees when there are people in the vicinity. Logging is dangerous work, and safety standards are required. They don't even fell trees with other loggers in the area. So these woods actions can theoretically protect a forest long enough for the authorities to arrive.

At Grizzly Creek, activists went out with maps and tried to show the loggers that they were cutting in an area that was illegal because it had endangered species living there. Pacific Lumber contended the harvest was part of an approved plan.

A logger, A. E. Ammons, became enraged. He was screaming at the activists, saying he wished he had his gun so he could just shoot them.

Thirty minutes later, David "Gypsy" Chain was dead. Ammons had begun cutting the trees in their direction. One hit really close. The activists regrouped, trying to decide if they were going to try one more time to reach out to the loggers or just leave because he was getting so angry. Then a tree came so close to them that they jumped up and started to run. A second tree came down, heading straight for them.

The forensics report shows that Gypsy did try to run, but friends with him said that he froze for a split second. That started rumors that he wanted to die.

"What was wrong with the guy?" people asked. "Why didn't he run?"

We're human, and fear is one of our human frailties. I imagine he panicked for a moment and stopped cold. Panic does that to people. It freezes our ability to move. He snapped out of it and started to run, but it was too late and the tree ended up taking his life.

When a tree is coming straight at you, it's impossible to see at first which direction it's going. Not until the tree is in the downward arc and is starting to crash through the smaller trees, when it has so much momentum that it's about to split the earth in two, do you realize it's coming right at you.

The tree hit so close that it covered the group with needles and broken twigs when it smashed into the ground. As people got up and started brushing themselves off, they looked around—and then started calling—for Gypsy. They didn't find him until Ammons went to buck up the tree and found Gypsy underneath it. The activists who were there say he dropped to his knees and began to cry. All the activists joined around with him. Together, they prayed for Gypsy, for the logger, and for themselves.

Only after I had finished my responsibilities as information hub, answering calls and networking to various people and the press, could I allow my own emotions to seep in. I started screaming "No!" louder and louder and faster and faster, as if I could reshape reality and bring him back. It was horrible that activists had to even be out there in the first place doing the jobs of government agencies; it was outrageous that I had to live in an ancient tree for nine months to try to protect it; it was beyond comprehension that a twenty-four-year-old had given up his life trying to get people to do the right thing.

Living in a tree made my senses so acute that I was keyed in to all the suffering of life, whether it was the animals or the trees or human beings. Without the distractions of society to numb me, those emotions were overwhelming. I had no immunity for this. So I was completely crushed when Gypsy died.

The first thing I did was call my mother.

"Mom, I need to pray with you," I said.

My mother is a very spiritual woman who believes, as I do, in the power of prayer. Prayers are the manifestations of our intentions. Together, we prayed for the healing and the love and the strength that it was going to take for many people to make it through what happened. We prayed to begin the healing process that we all have to go through when we lose loved ones, whether they be human or, for many of us, the forests and the Earth. We prayed that David's death not be in vain. And we prayed that I might find the strength to dig down deep and rediscover the love I would need to share with others in the face of this anger and violence.

Later, I wrote a letter to Charles Hurwitz, even though he won't take responsibility for Pacific Lumber—or the company

loggers—despite taking their profits. I didn't want anyone else on either side to die. I didn't expect him to answer my plea. But I knew I had to write anyway. Needless to say, he didn't respond.

Nothing prepared me emotionally for this tragedy. A few weeks later, near the end of September, people were grieving, people were still angry, people were still wandering around dazed and confused, figuring out what they were going to do. People felt anger, hate, frustration, sadness—and the desire to be like a wounded animal and strike out.

The line between holding someone responsible and striking out can be a fine one. We didn't want to spend all our energy attacking the man who dropped a tree on Gypsy, but we did want to state the facts. The fact is that he said he was going to kill someone, and then he did. When he dropped that tree, he didn't necessarily think it was really going to kill someone, but he did know people were there. He was trying to scare them, to mess with them. If he felled the tree and he didn't know that people were out there, that would be one thing. But he knew they were there, and he felled the tree anyway.

People wanted this man held accountable. They also saw that the government was refusing to do anything about it. After Gypsy was killed, the law enforcement agencies refused to cordon off the area as the scene of a crime, let alone the scene of a death.

"We can't call it a crime until we know all the facts," they said. But they wouldn't cordon it off so they could find the facts! The Humboldt County district attorney and sheriff were going to let Pacific Lumber go right back in and continue logging, pulling everything out, destroying all the evidence of what had happened.

So activists had to keep the loggers out, which is what law enforcement should have been doing. Using all sorts of fortresses and blockades and diversions, they took over the whole area, renamed it Gypsy Mountain in David's honor, and declared it a "free state," meaning that it was no longer usable for profit-making purposes.

The activists built a huge barricade at the gate. They dug ditches in the road and piled up big mounds of dirt, then they found an old junked car and towed it into the center of the road and locked themselves to that in an act of civil disobedience. For the loggers or police to come through, they would have to make their way through blockade after blockade, then they would have to remove the activists who were locked down on the car or locked to one another in the middle of the road.

A whole village of activists set up there. Residents from the local community would come by every day with food and blankets or whatever the activists needed. Hostile people would drive by and shoot guns in the direction of the activists at the gate. Once they threw a deer heart and hoof at them.

Energy started to wane after a while, which seems to be the history with frontline activism. Its intense nature makes it extremely hard to sustain. Then, at 5 A.M. on October 8, 1998, forty Humboldt County law enforcement officers in full riot gear stormed the blockade. The activists had let down their guard. Though the latter scrambled to lock down so the authorities couldn't simply remove them by force, they couldn't get in position quickly enough.

The police declared it an unlawful assembly and told people either to leave or be arrested. They hadn't done a thing to stop the violence against activists in the woods, but here they were in full force to prevent the activists from peaceably assembling. By the end of the day, the police had flushed out everybody who was not in a tree.

In the wee hours of the following morning, activists were able to sneak back into the area around the gate. The barricades were gone, so eight people turned themselves into a human barricade and locked themselves together across the road. Committed, passionate activists living their truth do not give up easily.

Meanwhile, the struggle to save Headwaters raged on. As part of the process for the Headwaters agreement, the government was supposed to allow for public input on Pacific Lumber's Habitat Conservation Plan. So the public was given a thirty-day comment period for a plan that had taken two years to develop and was over two thousand pages long. Hearings were scheduled in various locations, including one in Humboldt County on November 10.

I followed it closely, and it was quite an event. It went on all day and into the night and continued the following day. Two local noncommercial radio stations, KMUD and KHSU, carried it live, so I listened to it from Luna. People spoke on behalf of the timber industry, on behalf of the local communities, on behalf of indigenous peoples, *and* on behalf of endangered species such as coho salmon and marbled murrelets. Some simply spoke of the beauty and spirituality of the forest. Everyone from the timber side referred to Headwaters as an economic issue, even though this hearing was supposed to be about "habitat conservation" for fish and wildlife.

Pacific Lumber employees were given the day off so they could attend and speak. I felt sad hearing these people parrot the company line over and over, using the exact same words from the press releases that Pacific Lumber sent out to the media. But these people were not representing their own true interests. They hadn't bothered to ask themselves the obvious questions: If we cut at triple the rate now, what will happen a few years down the line? How will we make our house payments then? How will our children find jobs? Maybe they just hadn't taken the time to learn what the Habitat Conservation Plan and Sustained Yield Plan were really all about.

Their vision was too clouded by the god, as they know it, of Pacific Lumber. The government was offering to pay Charles Hurwitz huge amounts of money to prevent him from clear-cutting Headwaters, which we contended he couldn't legally do because of the Endangered Species Act. He wouldn't have to pay a cent of it to a single person in the mill or a single logger. He would take it all, and the workers would get nothing. But for some reason, they just didn't see that.

Suddenly, right in the middle of the hearings, it was announced that the California Department of Forestry had suspended and didn't renew Pacific Lumber's logging license after investigating the allegations that had propelled Gypsy and the others into action. Because Pacific Lumber was already on probation for prior violations of the Forest Practices Act (averaging close to one hundred violations per year, the California Department of Forestry revoked its license. A forestry department spokesperson termed the revocation "unprecedented" for such a major timber company. I and many others figured it was about time the company was held responsible for its blatant violations of the law.

"What do you think of Pacific Lumber having its license revoked?" a radio reporter asked one of the workers who had just spoken in the proceedings. "Are you worried about your job?"

"Not at all," he replied. "There's no reason to be concerned. Pacific Lumber is a good company. They've always taken care of their workers."

The very next week, he, along with hundreds of others, was handed his pink slip.

That weekend the workers held a rally. The signs carried slogans like: *CDF + Earth First! = Unemployment.* My heart went out to these people because they had bought Maxxam's propaganda, which pitted timber workers against environmentalists and therefore victim against victim. The truth is, both timber workers and environmentalists suffer because of Maxxam's policies. Pacific Lumber workers did not lose their jobs because of the California Department of Forestry, and they did not lose their jobs because of Earth First!. They lost their jobs because Pacific Lumber/Maxxam had complete disregard for the laws. If Maxxam had continued Pacific Lumber's tradition of logging in a more sustainable way—instead of instructing Pacific Lumber to operate under a policy of *take everything as fast as you can*—it would not have accumulated over three hundred violations in three years. It would not have had its license suspended. And the employees would have kept their jobs.

In the end, the employees were the ones to suffer from the suspension. Sometimes, the company found a swell loophole.

In some instances subcontractors were brought in to continue logging. Eitherway, Pacific Lumber employees were the ones affected.

Eitherway, it didn't last long. Within weeks, Pacific Lumber's license had been reinstated, same as it ever was. In effect, its license hadn't really been revoked at all, merely suspended for a slightly longer period of time than the usual slap on the wrist.

⸙

NOT ALL THOSE loggers and millworkers who had been laid off when Pacific Lumber's license was suspended could be hired by the subcontractors, who had their own pool of workers. So some of them traveled to Washington State to scab for Maxxam, where over three thousand workers for Kaiser Aluminum had just gone on strike.

Like Pacific Lumber, Kaiser Aluminum had once been a family owned and operated company before its shares became susceptible to a company like Maxxam. In both cases, Charles Hurwitz used junk bonds to make these killer investments. Then, to pay off the debt, he had to suck all the life he could out of each company before taking the profits and moving on.

When steel prices were down in the 1990s, Kaiser Aluminum's steelworkers had agreed to a cut in their pay because they believed in the long-term sustainability of their company. They were told in what they believed to be all good faith that their wages would go back up when prices rose again. Prices did indeed rise, Charles Hurwitz and Maxxam Corporation made a huge profit, yet the workers of Kaiser Aluminum were still making two dollars an hour less than other people in their area were making for comparable work, and with fewer benefits.

So the steelworkers went to Charles Hurwitz and Maxxam Corporation.

"Look, we kept our part of the bargain," the union said. "We allowed ourselves to make less money for a long time in order to help this company survive in an effort to invest in all of our futures, and now we're asking that the company give something back to us, the people who kept it going."

The company refused.

"But you've made a record profit," argued the steelworkers. "You made millions and millions and millions. You can afford to bring us up to standards. We're just asking for fair wages and fair compensation."

They also challenged Maxxam's practice of hiring out work to subcontractors who are not required to pay benefits and the same level of wages. With subcontractors, there's no job security.

The workers went to the table, but Maxxam and the head of Kaiser refused to negotiate a decent and reasonable solution. So the steelworkers' union went on strike. But instead of trying to settle the strike, the company locked them out. Though strikers opt to stop work, they remain employees of the company they are protesting against. When they're locked out, however, they're no longer employees. They're history.

To keep its factories going, Maxxam Corporation hired scab workers, some of whom had just been laid off from Pacific Lumber. Nobody stopped them.

When environmentalists heard what was going on with Kaiser Aluminum, they tried to get involved. Darryl Cherney and Mikal Jakubal were the first to go up there. Darryl started doing what Darryl does best, which is schmoozing with people and organizing. And Jakubal did what he does best: he went under cover as a worker, scabbing in order to find out what was really going on from the inside.

Thus began a strange alliance between environmentalists and labor. Striking steelworkers traveled down to Humboldt County to ask Pacific Lumber workers to support their strike instead of scabbing as well as to try to help the Pacific Lumber workers unionize.

When I looked at it, the potential of labor and environmentalists to coming together seemed big. Though we had some huge differences, we also shared a common goal: to fight the destructive power of giant corporations. Today's corporations have taken on more rights than the individual while assuming fewer of the responsibilities. The corporations have a pyramid structure, with the CEOs at the very top. These people make up a small fraction of the population of the planet, so small you can't even put them on a pie chart, yet the profit they make is what half the population of the entire world makes. The reason they're on top is that they're extracting the profit off the backs of hard workers and off the health of the land. The same forces that are destroying the environment are also destroying jobs and communities.

⟨~⟩

ALL THE PUBLIC controversy meant increasing media interest and, because of my peculiar visibility, the media interest in me only continued to increase. Even before Gypsy's death, the amount of traffic—journalists and others—was increasing in the tree, and so was the damage to the tree. People who aren't used to climbing trees don't know how to do it correctly and typically break branches. But each branch broken was one of my steps broken. They were breaking my house and my ability to live there. Not to mention, they were breaking Luna.

The stress on the tree bothered me, as did the wear and tear beginning to show on the platform and on the ropes that held it. Then Sawyer, one of the main people in the original crew of activists who had built the platform, called. "We're thinking about making a second platform on Luna so that you'll have a place for visitors and press," he said. "What do you think?"

Sawyer, a former carpenter with medium-length, reddish blond hair that looks as if someone had taken scissors and randomly chopped it, brought along DNH (short for Dave's Not Here), who had also worked as a carpenter for many years. Together, they decided that they would build a platform eighty feet below the first one, at a point below the ridgeline and lower than some of the surrounding trees. The tree splits in two there, then grows back together farther up. So there are a lot of really big, solid limbs on which the platform could be anchored. It would fit perfectly there.

They spent about three months on the ground designing and redesigning the platform, until they had it just so. Once they had built it to their satisfaction, they took it apart and, with the help of others, hiked it up the hill in pieces.

They hauled the skeleton framework up into the tree first, using all kinds of ropes and pulley systems. They set it by securing lines to different branches and then hooking those to metal anchors on the sides of the platform. Once they had leveled it out, they screwed on the salvage plywood flooring. Later, they hiked up the supports for the sailcloth that would constitute the walls and roof. By the time they were through, the finished product looked a lot like a covered wagon without wheels.

This lower platform was like the five-star hotel of tree-sitting. The rounded top made for a steeple effect, while the crisscrossing ropes that held it together formed a stained-glass design. So it looked almost like a church.

The peak of the arch being close to seven feet tall, I could actually stand up in it, something I was never able to do up top. There I'd have to go outside if I wanted to stretch to my full height.

Even better, on nice days I could roll the canvas walls up to turn it into an open, hanging platform. (All the branches, ropes, tarps, and duct tape that make up the higher platform's shelter make that impossible up there.) And because the new

"Still Up Here"

I sit here
swaying in the gusting wind
watching the winter rains
nourishing the thirsty land

I sit here
for over one year now
length of time has lost its meaning
to my ever-evolving mind

2/30/1999

walls and roof had clear panels, made from windsurfing sails, even when the canvas needed to be rolled back down on cold, wet, or windy days, I could still look out, and rays of light and sun could come in.

For a while, I moved back and forth between the two spaces. At first, I used the lower platform strictly for media interviews. Then in September, summer finally hit. Of course, it lasted only three weeks. But finally I experienced warmth. I was actually hot for a change.

In the top platform, with its lack of ventilation, however, I was too hot. In fact, I was sweltering. Tired of sweating, I realized how much cooler the lower platform would be. So I climbed down, rolled up the panels, and rolled out my sleeping bag. Gradually, I came to appreciate how much nicer it was down there: drier, cooler in hot weather, warmer when it was cold, and less windy. So that fall I moved down a floor and started settling in all over again.

Regardless of what Pacific Lumber may have hoped or thought, I wasn't going anywhere soon.

ELEVEN

A YEAR AND COUNTING

Winter. In a tree. Again.

On the night before its official start, the cold set in and it began to rain. I didn't sleep very well that night because my feet were so cold they hurt. I kept going in and out of consciousness; every time I'd wake up, I'd rub my legs and feet really hard and kick them around. It didn't help.

During the night, as if to warn me about what was in store, the rain turned to sleet and then to snow. That's when I started hearing the wildest sounds. A lot of branches and trees had been damaged during the logging, and with the weight of the snow they just broke. At night, the resulting sounds seemed ethereal. I'd wake up and hear a snap, and then a crash would echo in the stillness.

By morning, the snow had blown in through the cracks in my platform's sailcloth covering and blanketed everything, including my sleeping bag and me. When I sat up, the fine powdery snow slid right off me onto the platform. Nothing like life in a tree! I poked my head out of the back of the platform and watched my breath escape my lips.

Seeing the forest tinged like sugar cookies reawakened me to the fact that I was alive and reminded me of what a wonderful and precious gift life is. Part of me wished I could be out there playing. I tried making a couple of snowballs, but the snow was too dry and powdery to hold a shape.

The outside branches of Luna were laden with snow. Below, snow covered up a lot of the destruction with a glistening blanket. It was a fairyland.

The cold and snow had been difficult to take the first winter, because I was not prepared for either. As my second winter approached, however, people had started giving me many wonderful presents, including a bivouac, which is like a tent that wraps around my sleeping bag. So though the wind would blow rain or snow through all the cracks and crevices on my platform, when I slept I would burrow down into my bag like a caterpillar in a cocoon. I kept just a little space open for my face so I could breathe and tucked the rest of it around me to keep in my precious body heat. Aside from the physical comfort, the bivouac also gave me a huge burst spiritually and emotionally, because individuals I'd never met had been so kind as to send me this wonderful gear.

People donated all sorts of other things: a new, cold-weather sleeping bag (at last!), warmer clothes, lots of hats. Hats were hard to keep, even knit hats, because the wind would start howling and whip them right off my head. I lost four to the wind the first winter. They were like offerings to the four corners, north, south, east, and west. But, luckily for me, people kept sending me more.

I even managed to help my cold, painful feet. Toward the end of the first winter, activist and photographer Eric Slomanson had come up with these nice booties that had little air pockets to help trap the heat. He let me try them out; they were the first things ever that kept my feet warm. I bought them off him with some money I still had in my wallet from being on the ground and proceeded to wear them into oblivion.

I don't know whether it was my new clothing, my new gear, my new platform, or a year's worth of experience, but as the second winter approached, I felt a new sense of security. Cold and wet no longer seemed like terrible ordeals. They were just weather. Living in Luna had already taught me that one of the best ways to find balance is to go to the extremes. I was ready for whatever would come my way.

La Niña, as this winter would be known, didn't mess around. Hail followed sleet and snow. The hail in Luna was intense—like throwing pebbles on somebody's roof or the side of their house, *tadatadatadatada,* but magnified. The wind drove it through the cracks and crevices and into the platform. The frozen pellets would hit the branches and bounce inside, flying everywhere. I would be writing a letter to someone and suddenly have to put on my heavy rain gear to keep from getting battered.

Those were the moments when it would really hit me that I lived in a tree. I never totally forgot, but after a year living in Luna, it had become my way of life.

One whole year! It was hard to believe. Even more people braved the steepness of the hill to celebrate with me. For once, I was rendered speechless. People came from every direction imaginable. A celebratory rally and benefit was also held in southern Humboldt, at the Mateel Community Center in Redway. I got to listen to it live on the local community radio station KMUD, which made me feel closer to the event. I was even blessed with the opportunity to read my poem, "Offerings to Luna," over the phone to a sold-out crowd of eight hundred as well as to audiences on both the radio and the World Wide Web, accompanied by Mickey Hart and Planet Drum and Bob Weir on guitar.

I received one other very special gift: 475 stamps from some schoolkids in Wisconsin. Their teacher had heard about the tree-sit. After visiting our Web site, she had her fifth and sixth graders wrote me with questions about my stay in Luna. When I answered their letters, I added ideas about how they, too, could get involved.

Since the average American consumes—and disposes of—approximately thirty-seven tons of raw material a year, I suggested a science experiment where one half of the class would reuse all their scraps of paper and then throw away whatever they couldn't possibly use in a container. Meanwhile, the other half would automatically throw away their paper in a separate container without reusing it. At the end of one month, the two halves would compare how much paper they had.

Results varied with the size of the classes, but one found the half that didn't use their scraps has about six bags of paper trash per month more than the other half. Students could then multiply that by nine months and then by however many classrooms were in their school. They quickly saw how much waste, how much forest, most people just throw away.

I also gave them some practical ways to help, including getting their parents and their parents' business partners to donate money for tree planting, and starting organizations to save the forests in their school.

The students were so jazzed by my response that they wanted to help me. They figured the best way they could do that was to raise money for stamps so I could write similar letters to other children. They started out by talking to their principal and getting permission to broadcast a new little tidbit about the redwood forest over the public address system every day. Then they lined the school hallways with large billboards they had created about the redwood forests and rain forests. Every day they would add something new to their wall of information.

When it came time to collect stamps, they constructed a box out of local evergreens, put it on a little red wagon, and pulled it to each of the classrooms. The kids would take whatever stamps they had gathered that day and drop them in the box.

At the end of their allotted time, they had raised 475 stamps, which the teacher's daughter brought to me at my one-year party. She also brought me copies of the Save-the-Forest stamps they had designed.

That's young people for you. They are so ready to empower themselves and get involved. I loved it! I used those stamps they had raised to send letters to other children, just as they had wanted me to do.

Though I've never wanted to have children of my own, I've always cared very much about children and their welfare. At one point, I decided that I was going to work incredibly hard, invest my money, retire at age thirty-five, buy a nice plot of land, and begin adopting children seven years old or above, since that's when their chance for being adopted basically turns to zero.

Obviously my life has altered since then, but my care for children has not. That love helps renew my commitment to the forests. Even as we're destroying the environment, every day families are bringing new children into this world. What kind of world are they entering? Asthma in children is skyrocketing in cities like Los Angeles. Birth defects are on the rise as we dump more toxins into the environment. Young children are surrounded by violence, in the media and in real life.

Children, when they're born, are perfectly pure Creations. They are the source of Creation right there. From that moment on, the choices we make as a society affect their world. It hurts me when parents hit their kids, when children go hungry, when a child has a disease. Anything that hurts a child has always wrenched me apart. That's why children have been one of the driving forces of my activism, because we as a society have become so selfish that we don't think about how our actions are affecting children.

"What do you say to someone who can't relate to your experience?" I'm often asked.

"Go look into a child's eyes, and know that the simplest sacrifices you make today can be the greatest gift for their future," I tell them.

AS CHRISTMAS NEARED, I wondered: What greater gift could we offer our children than the protection of the last old-growth and wild places left on our planet? And what better way to remind us of that than a yearly period honoring the trees and the forests of the world? Since Luna has been called a beacon of hope, it hit me that we could light her up with beacons.

The first time I had made the grueling hike up the hill, the beacon that prior activists had hung had given me the hope I needed to make it the rest of the way. Now it was my turn to light up Luna with twenty-five battery-powered beacons that some wonderful supporters had donated. Climbing out on the branches, with fingers so cold they were no longer nimble, I hung them one at a time. Then, starting on the winter solstice, and continuing for about a week until the night after Christmas, I climbed around every evening at dusk to turn all of them on and every morning to turn them off. I'm told that as far as three miles away, and to motorists driving on Highway 101, they looked just like stars twinkling in Luna's branches.

Later, when the valley wasn't covered in fog at night, I would occasionally turn on a couple of beacons just in case someone driving by happened to look up. Mostly, however, I gave them away to the various people who have been important in sustaining the forest movement and the Luna tree-sit. If it weren't for the help of many different people on many different levels, the beacon that is Luna wouldn't have shone.

On New Year's, Stafford commemorated the second anniversary of the mud slide with a rally at the site of the slide. Efforts to convince Pacific Lumber to buy the residents' property for a fair price had floundered. Citing the distressed condition of the property, the company had offered prices well below the market value. And though Pacific Lumber had cleaned up the debris caused by the same mud slide that had torn through those homes (harvesting, in the process, millions of dollars' worth of logs), it had done nothing that would deter, or redirect, subsequent mud slides. If anything, the company's actions, which included erecting two berms and sediment ponds of doubtful stability since neither had been properly engineered, had increased that risk. Thirty-seven Stafford residents, most of whom still lived in terror of the mountain roaring down in the next rainstorm, had already decided to sue for damages.

AS 1999 SET IN, I began to lose track of the days. They just seemed to rise and set into one another, punctuated by phone interviews, letter writing, and visitors (invited or not). But one day in winter didn't need a calendar to mark itself forever in my brain.

Darryl Cherney, one of Earth First!'s organizers, called me one evening with an urgent message. He had gotten word that someone had fallen out of a tree on Gypsy Mountain. Could I get on the walkie-talkie (the only way to reach them) and confirm that?

All too soon, I learned that Bird, a gentle activist in his twenties, had fallen a hundred feet to the ground. Earlier that night, his fellow tree-sitter and partner, Lilly, heard a crack followed by a thud.

"Bird? Bird? Oh, my god, Bird!" she yelled in panic.

She hadn't calmed down much by the time I reached her on the walkie-talkie, which, because of low batteries, kept cutting in and out.

"Is he okay?" I tried to ask. "Is he speaking? Is he alive?"

Lilly was understandably in shock, and I could hear only every other word. It took me forever to figure out what was going on. I'd ask a question and try to get her to just say yes or no.

Finally, I managed to piece together what had happened and passed along the information to activists on the ground. Darryl Cherney was already in touch with the medics.

Between the two of us, we managed to relay their directives not to move him, to cover him with blankets, and to try to keep him talking until the medics got there.

The fog was too thick to get a helicopter in. When the medics finally did arrive on Gypsy Mountain, they couldn't get all the way up the road even on the four-wheelers they were forced to call in. So they had to hike in and hike out, carrying Bird. Eight hours elapsed from the time he fell to the time he got to the hospital. Had he suffered anything worse than a broken pelvis, I think he never would have survived.

I heard lots of stories about why he fell. Apparently, he was going onto a traverse line, and his carabiner was improperly connected to the climbing pulley he was using. When the climbing pulley fell, so did he. The branch that he hit on the way down—the crack that Lilly heard—must have helped to break his fall.

I was so thankful that when his wings failed, the wings of angels were watching out for him. Something must have flapped for him a little bit, since he fell that far and did not die.

Even before Bird fell, people would often ask me whether I ever came close to falling. The answer was always no.

"Is that because you're so comfortable?"

"No, it's because I'm so careful."

It would have been easy to feel too comfortable in Luna. Connecting to her in such a strong way was a heady experience. But when we become too comfortable, we make careless mistakes. And at a hundred to a hundred eighty feet off the ground, a fall, or even an accident, could kill me. So even when I slept, my senses remained attuned, because a creak or a groan could have meant that something was breaking that my life might literally depend on.

I couldn't afford to ever really relax, because I couldn't afford to make a mistake. And not just on the physical front. I had to be on guard spiritually as well. Carrying so many people's hopes for the forest movement was a huge responsibility that I took very seriously. With so many struggles and with so much pressure, it wasn't easy to avoid thoughts about rejoining the world. In a poem entitled "Down," I summed up my quandary:

If people's hopes are placed on me
and I come down
do their hopes come down with me?

The fact that my actions, increasingly spotlighted, affected so many more than me meant that I had to weigh the consequences of each word and deed. Everything we do ripples out and affects other people's lives. But what I did and said affected people's perceptions about the forest, environmentalism, and direct action. If I had made even the smallest mistake, the timber industry (and even corporate government) would have pounced and exploited it as much as they possibly could. And by discrediting me, they would have stripped other activists of their credibility. That was a huge weight.

As I settled into the rhythm of my second year, I felt

drained, exhausted from my nerves being on edge all the time. "Is this forever?" I wondered.

I wanted a shower so badly I could taste it. I could feel the hot water pouring over my body, into my pores and through my hair, which was in such need of cleaning. I wanted to be able to wake up at night and not have to pee over a bucket or a funnel if I needed to go to the bathroom. I wanted to sleep a full night through, not wondering if I would have to grab on tight when the wind picked up or whether I'd survive to hear the wind pick up the night after that.

Yet each time I'd start to feel that the fire inside me was just too weak to burn any longer and that I couldn't face another day, the great spirits of the universe would send something to fan those flames and burst them back into the bonfire I needed to renew my strength. Sometimes it would be a call from a friend. Other times it would involve a prayer being answered more quickly that I could have thought possible. Occasionally, even Pacific Lumber, however unwittingly, contributed to my resurgence of spirit.

On February 8, 1999, the same week that a wrongful-death suit was filed against the company for the loss of David Gypsy Chain's life, Pacific Lumber guards returned to post an eviction notice. The notice, which they hoped would protect them from future liability claims, said they were concerned with our safety and the safety of their employees, that we were engaged in an illegal activity, and that if we refused to leave their land, we would be subject to arrest under specific laws and codes, which they cited. It was all done very formally. They nailed a copy on Luna and the other trees with active tree-sits, and they published it in the legal notices of the *Times-Standard*.

I climbed out of the platform when I heard the *tap, tap* and felt the *thumk, thumk* of the hammer.

"We're posting a notice. Come down and read it," the men said when I asked what they were doing.

"You guys know I haven't been down in over a year," I retorted, adding that I didn't think that it was very fair for them to post an eviction notice on Luna, since she couldn't pick up and move.

"I'm sure she would move if she could," I concluded. "The neighborhood has gone to pot since Charles Hurwitz moved in."

The pressure to come down wasn't limited to Pacific Lumber and Maxxam. A lot of people within the movement felt that I should leave Luna and help bring an important spotlight on other environmental battles. But the pressure other people put on me was nothing compared to the pressure I put on myself regarding this action. I cared about the forest, this planet, our world like I'd never cared about anything before in my life. So the question of whether to come down or not took on a whole new level of responsibility and a whole new level of reality. It even entered my dreams. I would toss and turn at night, anxious and confused about what I was supposed to do.

Under pressure, I have trouble hearing the guidance I live my life by. While I take other people's thoughts and concerns into account—I've never pretended to be a know-it-all—I get my ultimate guidance from prayer. That's why I pray every morning and every night. But though I prayed really hard about coming down, I remained trapped in this funky limbo state, torn about what to do next. As much as I would have loved to go down, I knew that that event would be a once-in-a-lifetime opportunity to get my message out.

I was feeling pulled in every direction. Part of the world wanted me up, part of the world wanted me down, another part of the world wanted me dead, and I had to try to figure out what to do. It was overwhelming, especially when I was trying to write a billion letters, conduct so many interviews, and fulfill speaking engagements.

At that point I reminded myself, as I have at many times in my life, to take my time and remember to breathe. This world is so fast, and there's so much pressure to move now, move quickly. But I knew that if I wasn't feeling clarity, I had to take the time to let the right thing happen. I couldn't let other people sway me just because I was unsure. That was part of the lesson that Luna had taught me: to be still and listen, even in the chaos of my life.

I knew prayer had taken me to the Lost Coast, prayer is what guided me to the redwood forest, and prayer is what led me to this tree and up this tree. Prayer is what had given me the strength to continue all this time. And someday, prayer would help guide me down.

Snowfall in Luna

TWELVE

INSPIRATION

The good days in Luna were fantastic, to the point where laughter split my lips, over and over and over without control. The bad days were so bad that nothing but tears fell out of my eyes and my heart. The bad days involved fierce wind, rain, sleet, and hail, chain saws going in the distance, the yarder with its incessant beeping, beeping, beeping, picking up logs, and helicopters raping the hillside.

Bad days lay ahead.

On March 4, 1999, after Pacific Lumber regained its timber operator's license, the weather cleared. I knew the helicopters couldn't be far behind. The thought of their intense power, shaking and breaking everything, filled me with dread.

The company argues that helicopters represent a more gentle form of logging. They may be more gentle on the ground, but you should see what they do to the canopy. Their three-hundred-mile-an-hour updrafts take a tree and twist it up like a rubber band. Then, when the helicopter moves its position and changes the updraft, that tree unwinds. And when it unwinds, branches break off and the canopy goes flying. Those trees left standing look like they've gotten a bad haircut. In time, many die. The damage they've sustained leaves them unable to withstand the weight of the snow or the force of the wind, and—*wham!*—they fall over or break.

Though the helicopter didn't try to remove the fallen trees close to Luna, it did return to take out the rest of what's referred to as slash, which is actually broken forest. When loggers cut an old-growth tree, particularly on a steep slope like Luna's, they often prepare a bed of smaller trees to cushion its fall so it won't shatter when it hits the ground. Now it was time to pull out those beds of trees as well as the others that were unintentionally broken and smashed by the old-growth trees during their fall, and still more that were ripped from the ground by the first round of helicopters. They were there to carry off as much of the hillside as they could get away with.

Once again Luna was in a war zone. A large Columbia helicopter, its twin propellers sounding like machine guns, would hover overhead as the ground crew in bright orange clothing attached the trees to a dangling cable. About every two hours the helicopter would have to go refuel—a blissful moment of peace. But before I knew it, I would hear the blades off in the distance and then it was time to batten down the hatches, get ready for another round.

The helicopters didn't hover right over my head this time, so I did not feel my life was in danger from the updraft of the helicopter or flying tree limbs. But that didn't help other trees or the wildlife. Those three-hundred-mile-an-hour winds took their toll once again on the remnants of this forest.

Just as they had the first time the helicopters appeared, all the animals left. Not a living creature remained. No deer, no squirrels, nothing. Birds wouldn't fly over this area, not even vultures and ravens.

The helicopters came back day after day for weeks, every time there was a break in the weather. I started hearing those rotors in my sleep. My ears rang with a constant buzzing—*eeeeeeeeee*—for weeks after they'd gone.

Their absence due to stormy conditions was small consolation.

THE FIRST WINTER, I had dealt with the nonstop rains and wind that El Niño brought but only one snowstorm. This winter, I would be covered with snow eight times. The first time around, my job was to hang on so I wouldn't blow away. This time, my task was to keep from freezing to death.

"How are you staying warm?" Geraldine from KHSU radio asked me during one of the many back-to-back storms.

"I'm wearing two pairs of socks and some booties, two thermal pants and some wool pants and then ski pants over that, two thermal shirts, a wool sweater, two windbreakers, a raincoat, gloves, and two hats. I'm getting close to being as wide as I am tall, but it works." Though I had started eating mostly raw fruits, vegetables, and salads by that point, I would comfort myself with hot tea that I'd brew on my single-burner camping stove. That was my splurge.

When visibility was good, I watched the creek across from me feeding mud and debris from the clear-cut into the Eel River, turning it black. In the words of a nearby resident, "When you log steep hillsides and burn them, it will end up running into the creeks and filling them with silt." Even in the midst of these storms, there was so much to remind me about the work that still needed to be done if we were to stop—and reverse—the destructive cycle.

March started badly. The Headwaters deal needed to be signed by the first at midnight, or $250 million in federal funds would have gone poof! and turned to smoke.

As the deadline approached, the story was all over the news. I didn't have much time to listen to the radio, however, because I was too busy talking to those same reporters. For close to a week, I was lucky to get two hours' sleep a night. On the night of the deadline I ended up passing out from sheer exhaustion sometime between 10:00 and 11:00.

At approximately 12:20 A.M., I was awakened by a phone call from a reporter.

"I guess you heard the deal went through," she said.

"I'm hearing it from you just now, but I knew it a long time ago," I responded groggily.

Everyone who had done any research on Charles Hurwitz knew the deal would go through because Charles Hurwitz is a very shrewd man. On many occasions, including the takeover of Pacific Lumber, he has pulled the same tactics: throwing his corporate weight around and then bluffing all the way to the last minute to extract as much as he can from every single deal.

In the Headwaters case, he bullied people by holding the forest hostage and saying, "You give me what I want, or else." Some improvements—including stipulations for more protection of steep slopes and Class 3 streams (those that run only in the wintertime) in order to save the coho salmon—were made to the original Habitat Conservation Plan. But Hurwitz managed to keep those to a minimum.

In the final hours before the deadline, despite everything that was being done to help Charles Hurwitz make $485 million, the latter claimed that he still would not have enough money to pay off his bondholders. So the deal stalled. In keeping with our "political" process, government and agency officials once again retreated behind closed doors with representatives of Pacific Lumber and Maxxam Corporation. By arbitrarily changing the midnight deadline from Eastern Standard Time to Pacific Standard Time, they bought themselves an extra three hours and miraculously managed to find an extra forty-five million board feet per year available to harvest. That constituted a 30 percent increase above the level that all the scientists working on the Habitat Conservation Plan and Sustained Yield Plan had deemed acceptable. What magicians! By altering time itself, they had succeeded in conjuring up tens of millions of board feet of lumber out of thin air.

This was not the first time politicians had stopped the clock for the Headwaters deal. Earlier, the state senate was supposed to have its appropriations bill completed by midnight. So just before midnight they literally stopped the clock in the senate chambers. Technically, by the clock, they pushed the bill through.

"Extortion!" screamed an onlooker seated in the balcony when they finally finished, hours later. No one word could have been more appropriate. The forest had been made to pay.

The Headwaters deal had managed to protect groves that were here long before Columbus set foot onto this continent. But it was a bittersweet pill. The cost of that protection—the eventual desecration and devastation and destruction of tens of thousands of acres—made me cry. It also furthered my resolve to continue on with the struggle.

Charles Hurwitz should never have been paid a penny for Headwaters. By my account, Charles Hurwitz didn't even "own" Headwaters: Hadn't the taxpayers already paid $1.6 billion dollars to bailout United Savings of Texas? And even if he did own it, the taxpayers, by some accounts, paid three times what the land was worth if environmental laws prohibiting clear-cutting were taken into account. Meanwhile, Native peoples had to sit by with hands tied by so-called leaders as the Earth they once showered with love and respect was basically stolen from them once again.

The politicians couldn't praise themselves enough over the agreement. I was outraged. The day after the Headwaters deal was closed, I was part of a press conference that was recorded and broadcast over the Pacifica Network News:

> I am still up here in this tree in an amazing windstorm because this deal is going to cause more harm than good. President Clinton says he is proud to be a part of leaving this legacy behind. I would like to ask Mr. Clinton to come sit in this tree, where I have been sitting for over a year, and look at the legacy of mud slides, look at the legacy of clear-cuts, look at the legacy of the Eel River that turns darker and darker with the sediment from the clear-cuts as every storm hits the area. I'd like for Mr. Clinton to see the legacy of these destructive patterns of Pacific Lumber, multiplied by the thousands of acres that he has just made it legal to destroy. And I would like to ask him: "What is there to be proud of in that?"

On that same day I told Geraldine from KHSU:

> Everyone within the government is touting it as a wonderful deal, and many of them are saying they're glad to know that they've helped bring about peace, much-needed peace for these communities. I wish they had done that; we all would love to see peace here, but this deal won't bring that peace.
>
> It's not that we thrive on the constant struggle, but we can't accept what they've given us because it's not

enough. It's not enough not because we want it all, but because all that's left is so little.

To know that every step of this deal has been a compromise makes me shudder. We're talking about species of plants and animals, we're talking about people's lives that are in critical condition, and to compromise on a critical condition is death. That's why I'm still up here, that's why we're still committed to working toward real and necessary protection.

The Headwaters deal made it clear that our real work had just begun. We had to gain protection for slopes that are too steep to be harvested. We had to put an end to clear-cutting and the subsequent use of napalm, diesel fuel, and herbicides. We had to make people see the great harm caused by industry replacing forests with "tree farms."

Many timber companies, including Pacific Lumber, say, "We no longer practice clear-cutting, or we practice it only rarely." Yet from where I sat, for miles in every direction, I could see barren, burned swaths of destroyed ground, with a few dying trees left standing in the middle. A clear-cut by any other name, such as "even-aged management" or "variable retention," is still a clear-cut.

❧

DESPITE MY PUBLIC FURY over the Headwaters deal, my conversations with John Campbell continued. We were speaking about once a month, and he'd always try to get into these debates on the environment versus the economy and what wonderful stewards of the land Pacific Lumber was.

"You know, we're never going to see eye to eye because I don't get money for what I do, while you do," I told him. "If you weren't getting paid, Mr. Campbell, you wouldn't be doing what you're doing."

He just laughed, and then we changed the subject.

After much small talk, Campbell gradually started to open up more on a personal level. I asked him about his past and how he ended up in the United States, about his marriage and his family. He talked about his son, who was graduating from high school. If the latter got into a good college, Campbell had promised to take him on a trip wherever he wanted to go. So they were headed to Australia, John Campbell's homeland.

Soon we were talking every couple of weeks on an increasingly personal level.

"You know, I sent your mother a letter, and I never heard back from her," he said at one point.

The letter included the suggestion that my mom counsel me to come down since what I was doing was so unsafe.

"That's because your letter was a bunch of your company lies, and she wasn't interested in hearing it," I retorted. "She's a busy preacher's wife. She spends a lot of time caring for the sick and attending to the families of the dying and helping people arrange their weddings and all the other things that a preacher's wife does. She's not going to take the time to answer your company rhetoric. If you had sent her a letter that was substantial and human, she would have responded."

He just laughed. When the truth hits too close to home, Campbell laughs and changes the subject.

We still weren't getting very far with terms for a resolution, but at least we were talking. When two human beings come together, the communication has to start somewhere. Basically, I was saying to John Campbell, "Hey, look, I'm a real person up here in this tree." I was not making demands, I was asking what it would take for him to protect Luna and her surroundings.

I was like water wearing away at the stone. Water acts differently than a hammer and chisel, which chip away at something. I was just a constant presence that sooner or later would be heard. Not because I'd pounded in the message, but because I was always there.

For a while, Campbell sounded like a computer printout from a PR firm. Over time, however, I could hear the shift in his voice and his words. On warm days, he would tease me, asking if I didn't want to come down and go for a swim in the river's swimming holes. Then he offered to take me out for pizza and cappuccino. I guess he figured those were the things that a twenty-five-year-old living in a tree would miss most. It became our long-standing joke. From then on we managed to integrate cappuccino—or beer on the really hot days—into every conversation.

"That cappuccino is just waiting. It's getting cold," Campbell would say.

We'd gotten a lot of mileage out of cappuccino and beer.

In April, he switched tactics again.

"I'll tell you what. I'll meet you somewhere near the base of the hill, where no one will know that you've come down. The tree will be safe; I'm not going to send anyone up there. Nobody has to know that we've met," he suggested. "We'll talk, and if you don't like what I have to say, you can climb back up into your tree."

I felt torn. He sounded like he was reaching the point where he might be willing to protect Luna and the grove. Coming down might be the concession needed to make that happen. But another part of me warned that as soon as I came down, so would the spotlight, along with Pacific Lumber's incentive to work with me.

Around the same time that Campbell was trying to convince me to trust him and come down, a logger showed up at the base of the tree. He started hurling insults, but eventually we began to talk.

"You know, Julia, Campbell's never going to negotiate with you while you're in the tree. It's as if someone was sitting on your front porch and wouldn't leave until you had given them half of your house."

"That's just what Mary Bullwinkel always says," I answered, laughing. "Don't you have anything original to say?"

The logger laughed, too.

"It's the truth, though," he said. "He's never gonna negotiate while you're up there. I wouldn't, either. You don't have a right to be doing what you're doing."

I knew, of course, that Pacific Lumber/Maxxam were the ones who didn't have a right to be doing what they were doing. But I wondered about the pressures that Campbell might be experiencing, if that was the common sentiment. Still, I had given my word that my feet wouldn't touch the ground until I had done everything I could to save Luna. To take it to this level and then abandon the tree would be to violate my word.

"I can't come down," I said to Campbell over the phone. "But why don't you drive up the logging road? From where you park on the clearing, it's just a half-minute walk to the base of the tree. I'll come down."

"You'll come down to the ground?"

"No, but I'll come down to a branch low enough so that we can talk."

To my utter amazement, he agreed. If I had been sitting near the edge of the platform, I would have gone tumbling off.

"I'm not coming into the tree," he warned.

I said I understood but told him he ought to consider it someday because it is truly an amazing experience. He laughed and said he'd just stick to boyhood tree-climbing memories. At his request, I also promised that I wouldn't turn his visit into a media circus.

"And you're not going to swing down and swoop me up into the tree, right?"

I laughed.

"You've been watching too much Tarzan and Jane."

Two days later, Campbell showed up on schedule. When he honked his horn to signal his arrival, I climbed to the top of Luna and waved hello. We started to talk across the two-hundred-foot divide that separates the top of Luna from the clearing, located at about the same elevation.

"I should charge you double rent," he bantered, noticing that I now had two platforms instead of one.

As Campbell walked to the base of the tree, I prepared to rappel down to meet him. I wanted to go low enough for him to see my face, to see me as a flesh-and-blood human being. I hadn't put on my harness in months, however, so it took me longer than I expected.

"I brought you a six-pack of Pepsi as a gift. I hope you don't mind, but I drank one while I was waiting," Campbell said when I finally made it down to one of the lower branches.

I laughed, not having the heart to tell him that I don't drink Pepsi. I was thankful for the gift. When I was a young girl, my parents taught me to always accept gifts graciously, even if I didn't want or like them, since in the giving of an object, people are giving of themselves.

Then I lowered down my gift to him—a little stuff sack containing a crystal from Mount Ida, a powerful energy-vortex mountain in Arkansas.

It was so funny. He was trying to bring me what he figured I would most appreciate, living in a tree. And I was trying to offer something that would open his heart and his spirit. Pepsi versus crystals—gifts representing the two completely different worlds we came from. And yet, we were managing to communicate.

"So, what do you envision for protecting the area?" I asked as I strapped the five-pack of Pepsi to my waist.

"I see us protecting Luna."

He called the tree *Luna!*

"This is fantastic," I thought. "We've really gotten through to him."

"We'll put a plaque up here saying that Luna is protected forever. You can come down, and we will finish this Timber Harvest Plan according to California Department of Forestry rules. And you can come down and we can have that pizza and cappuccino."

"That sounds fantastic," I replied.

"Okay. I'm going to go back, and we'll write something to make sure this happens. Then we'll go from there."

He started to leave.

"John, I really think that we need to protect a little more than Luna. I don't want Luna to just be standing in the middle of a mud slide, because then she won't stand for long. Luna needs a family around her to keep her standing."

"Well, we'll work that out," he promised.

❧

IN THE MEANTIME, however, business continued as usual.

Clear-cutting involves not just the cutting of trees but the entire management system that follows. When companies clear-cut, they destroy nature's way of dealing with fire and invading plants. So they set fire to all the slash—twigs, brush, and branches—in what they call a controlled burn. They claim they're mimicking nature's fires, but their process destroys all life in the area, including the nutrients and microorganisms in the soil that are vital for a healthy forest. Nature's fires act as a form of renewal; Pacific Lumber's fires do nothing but destroy. When the fires cool, it's like someone has painted over all the different shades of green, from light neon to deep, deep dark green, with a brush dipped in monochromatic brown.

Once they have cut and burned and there is no longer any forest canopy, invading plants creep in. Many of these grow faster than the trees of Pacific Lumber's tree farms. Since these weeds, as they call them, would slow the growth of their merchantable timber, they spray poisonous herbicides onto the land, using diesel fuel as a carrying agent. (Our government is allowing them to do this; in fact, as part of the Headwaters deal, it will be legal for the next fifty years for Pacific Lumber to dump forty gallons of diesel fuel per acre on their clear-cuts.) By this time, they've not only raped nutrients from the soil, they've also poisoned the land, the water, the animals, the people.

Last but not least, because they've depleted the soil of so many of its nutrients, they must come in and dump chemical fertilizers on the land. They follow this destruction and devastation by the planting of two trees for every one they've cut down, producing miles of diversity-free tree farms, all the while patting themselves on the back saying, "Look what good stewards of the land we are."

⤙⤚

A FEW WEEKS after the Headwaters deal was signed, I witnessed firsthand what this meant for the forests surrounding Luna. After carefully leaving a few trees standing in the middle of the logged area, so it wouldn't be considered a clear-cut, they burned the whole thing with the help of a helicopter. It was just

This is what napalming looks like from Luna. A helicopter flies in the middle of this photograph, napalming the earth after a clearcut.

a tiny machine, nothing compared to the big Columbia they use for logging. It trailed a long cable with a large metal canister suspended at the end—an angry wasp with a stinger.

Shooting liquid fire—also known as napalm—from the tip of the canister, the helicopter sprayed various spots until the whole area billowed with smoke and flame. When it was all done the few trees they had left standing were completely fried, along with everything else.

Just over the ridge and in every direction—west of me, south of me, and east of me—for miles and miles, fires burned. At night, I could see the hills alight with glowing embers. There were so many of them it almost looked like a lava flow.

The fires lasted six days. The smoke, nasty and filled with the smell of diesel fuel, lasted long after. I'd never been subjected to that kind of torture before. When you live on the ground you can go into a structure with sealed doors and sealed windows and get away. Living in a tree, I just had to sit through it with only a wet bandanna as a barrier. My eyes swelled almost completely shut. My nose dried out and started bleeding uncontrollably. My strangled sinuses whirled. Every breath I took into my body, hour after hour, day after day, was choked with smoke. My throat burned. So did my lungs. I cried and prayed and cried some more. I finally began to pray, "Okay, God, either kill me or make this stop, one or the other."

The physical torture was compounded by the psychic misery I felt. Helicopters napalming a countryside, just like in

Vietnam—this was what the Headwaters deal meant for the rest of the land still owned by Pacific Lumber.

"Why are you still up there?" reporters would ask me once the Headwaters deal had been signed.

The answer seemed obvious.

"I am still up here because the unsustainable practices of the Pacific Lumber Company must be changed," I'd tell them. "Luna is still not saved, this hillside is still in critical condition, the lives of the people down in Stafford are still threatened, and all of this is indicative of the thousands of acres that have been sacrificed to this deal. I remain in Luna to let the world know that the forests here are still under attack."

MY WORDS WERE strong. My spirit was not. Then a group of striking steelworkers, who also had been suffering, came to Luna. During the one-year celebration, in their first public show of support for what I was doing in the tree, a group of them had hiked up to Luna with the banner picturing steelworkers on one side, the forest on the other, and an oversized Hurwitz in the middle stuffing a huge pile of gold coins into his arms. Now, as Earth Day neared, David Foster, one of the leaders of the United Steelworkers of America, visited me personally. He had come to Humboldt to speak at a rally, and he decided he wanted to come up into the tree. Even though he had a tight schedule, he made sure that a visit was included in his day. Foster was absolutely determined to climb into the tree on his own. This is not an easy thing. As I've mentioned, climbing into Luna is very, very difficult—it takes rhythm and a lot of strength in just a few muscles, because the same ones are used over and over and over again. And it takes time. Some people who are very good at it, like Spruce, can do it in five or six minutes, but for the average person doing it the first time, it takes anywhere from twenty minutes to two hours. It looked like David Foster was going to be one of those two-hour ones, and he didn't have that much time. So Spruce set up a pulley system. He climbed up to a strong branch, then rappelled back down the other side and hooked a cable up to David Foster. Spruce kept him hooked on two lines, one the safety, the other the rope that would lift him into the tree.

Then these big, huge, gargantuan steelworkers started hauling the line, and David Foster flew up. He made it to the platform in about a minute and a half, the closest thing to human flight other than shooting people out of canons. When he got to the platform, his eyes were about as big as his head. It's a little overwhelming to be in a tree at all no less this high up in a tree, and he was pulled up so quickly.

We had just a few minutes together, maybe ten or fifteen, before he had to go back down. We talked a little bit. I asked him about himself and his history and about his feelings about the coming together of the labor and environmentalists. He said that he really wanted to come up in person and say thank you, because the action that I've taken apparently has created a lot of inspiration for the striking steelworkers. He said that every time he's heard a worker grumbling and getting despondent, he's heard somebody else say, "Think about that woman up in the tree. If she can live up in a tree for over a year, you can keep on with the strike and help make the world a better place, too." And then the grumbling stopped. He said it's really tough for these guys who are used to having a steady paycheck, and now they don't have one. They're struggling against this huge machine of a corporation that doesn't seem to care, and they get really depressed and despondent. He said, "Every time, the thought of you up here has picked them up." I felt blessed and honored to be able to give strength to someone. And I am more than gratified to be a part of a union of two such seemingly different movements, labor and environmentalists. Little did I know how these small seeds, along with others planted by other caring people, would turn into the beginnings of a new movement that combine to fight together for our planet and our livelihoods.

The steelworkers played a role in another wonderful event that came on Earth Day itself. The night before I hadn't had an hour's sleep because there was a raging windstorm. I slept maybe five minutes here and ten minutes there, so I halfway loopy just from lack of sleep. But in the middle of a day full of interviews (I guess my story was pretty good for Earth Day coverage), I heard there was going to be a "surprise" visit from Bonnie Raitt, Joan Baez, and a bunch of steelworkers. Now, it's pretty hard to surprise me—I do live in a tree, after all, and can see for miles—but because they had tried to surprise me, the person who usually sets up the gear in the tree, my superman Spruce, was not around. He was off in Georgia doing an animal rights action. So Michael, my other main support person, was supposed to do it, but he hadn't been trained to set up all the ropes and pulleys. Finally, Sawyer came up, but he hadn't set up this complicated system before either, and he didn't have time to test it out.

The idea was to set up a pulley system with a huge rope, just as we did with David Foster, and have six or eight steelworkers with great, huge muscles on one end of the rope, with a famous musician attached to the other end. The idea was that they would give a heave and a hoe, and up she would fly. It was supposed to be a human-powered elevator ride, but it didn't work exactly as planned.

Bonnie was the guinea pig and went up first. The wind was howling, and it blew the rope onto the wrong side of a branch. It was too difficult to maneuver around the branch while they were pulling her, so they had to lower her back down, and I had to get the rope untangled from the branch. They pulled her up again, but just before she got to the top, we found that the two ropes— the main line and the safety—were set up on two different sides of the same limb. So she was about to be split in two! I had to do all this crazy maneuvering—hooking and unhooking, a safety here and a re-safety there—to get her around the ropes and under the branch and hooked back up. It was nuts!

Finally, we got Bonnie up and safely on the platform. I went back out and undid one of the lines and pulled it all the way up and lowered it back down on the other side of the branch. Then we got Joan up without too much trouble. I know people imagine living in a tree is peaceful, but not that day! Two film makers who were with them, Jan and Duff, wanted to come up right away, but Bonnie and Joan said they wanted to be up with me alone for a while. But after a few minutes, the walkie-talkie started squawking first with one person, then another. So I just turned the volume off without even telling anyone, because Bonnie was trying to tell a story and I wanted to hear what she had to say. Meanwhile, my pager was going off, so I turned my pager off, too. And I turned off the ringer on my phone, too. Life in a tree!

Finally, we got a few moments of peace and Bonnie and Joan told me their stories, and I shared some of my poetry. Bonnie and Joan sang some songs (they would later do a benefit for the Trees Foundation two nights before my second anniversary in the tree), and I felt humbled by such an honor to have these beautiful, brilliant women visit me in Luna. They are the most powerful examples of people who have fully integrated their creativity and their passion and their activism into themselves. Bonnie's personality matches her hair—she has that fiery red hair and a fiery personality, lots of sarcasm and humor. Joan Baez is more reserved, but she's got this impish side of her that finds its way through despite her reserve.

Bonnie presented me with her Bammy award, which she had just received for lifetime achievement. She dedicated it to me and actually gave it to me up there on the platform. I was so amazed that she would do that that I was speechless with gratitude. Unfortunately, because it's a big piece of Tiffany glass, I had to send it back down to the ground. Bonnie said, "Why don't you keep your fine Tiffany up here for the next time you entertain?" But it's this pointed piece of thin glass, and things tend to fall between the cracks up here. It was a weapon ready to annihilate someone, an accident waiting to happen.

When it came time to leave, neither woman wanted to go. But they had to. There was a moment of panic when Bonnie realized that she had to jump off a platform a hundred feet in the air and rappel down, but she did it. Waiting at the bottom were those supportive steelworkers who bore witness to this funny mix of constituencies. Although these groups would meet again in Seattle in December 1999, this was a powerful moment of alliance. That was Earth Day, 1999. I felt truly blessed by the experience, but I was also really thankful when the final person was down from the tree.

⟋⟍

BY THE MIDDLE of April 1999, just a few days after his visit, Campbell and I began discussing the protection of Luna in concrete terms. We went back and forth trying to figure out what would be mutually acceptable. I mentioned that the helicopters weren't supposed to come within two hundred feet of the tree, a distance that, scientifically, represented a good buffer zone.

"How about protecting a two-hundred-foot buffer around Luna?" I proposed.

"What about the downed timber in there?" countered Campbell. "It's already dead. There's no point in leaving it to rot."

That was really hard for me. Coming in with heavy equipment or a helicopter would change the landscape around Luna, at least for a while. The underbrush had begun to grow back over the fallen timber, and that rotting wood had a purpose. It's vital that trees fall to the ground to put nutrients back into the soil. Large, woody debris provides habitat for helpful fungi and mycorrhizae, and it also helps stabilize the hillside.

Things were looking up.

But that was the key negotiating point. I knew that sooner or later we would reach a point where I would have to swallow something distasteful, and this was it.

I debated with Campbell.

I debated with myself.

In return for permanently protecting an area slated for destruction, I wouldn't be selling out another forest, as they did with the Headwaters deal. I wouldn't be approving an arrangement that would poison the land or hurt anything else. All I had to do was come down.

Finally I agreed. They could take out what they had already cut down. In return, we would get Luna and a two-hundred-foot buffer of what remained. Actually, quite a bit of it was left, including several large trees, two of which had been marked to be cut down.

We had a deal. Or so it seemed.

THIRTEEN

THE DEAL — MAYBE

A year and a half in a tree! I found it difficult to believe I was still there. And yet I had lived in Luna so long I could hardly imagine living anywhere else. The tree had become part of me, or I her. I had grown a thick new muscle on the outer sides of my feet from gripping as I climbed and wrapping them around branches. My hands had also become a lot more muscular; their cracks from the weathering of my skin reminded me of Luna's swirling patterns. My fingers were stained brown from the bark and green from the lichen. Bits of Luna had been ground underneath my fingernails, while sap, with its embedded bits of bark and duff, speckled my arms and hands and feet. People even said that I smelled sweet, like a redwood.

The daily gifts that went along with living in Luna helped me to continue giving everything I had in order to save her. Like the day a squirrel jumped on my knees while I spoke into a tape recorder, lifted its paws up in the air as it stood on its hind legs, and poked its nose up at me. Or watching the sun rise, orange and red and peach and gold, and shoot across the fog in the valley. Or having a black bear, searching for blackberries behind Luna, do a double take upon seeing me in the tree.

Frankly, I don't think Campbell could quite believe that I was in the tree, either. Not after all this time. But that seemed about to change.

Or would it?

A month and a half after John Campbell and I had agreed on the basic outline for a settlement, I was still living in Luna. Our conversations—and agreements—about her protection had escalated into a world of lawyers and congressmen.

The agreement, which started out as a couple of pages, mushroomed into fifteen. It covered all the possible loopholes, like fire or slides, that the company might use to "salvage logs." On my end, I had to agree not to climb into their trees again or even to set foot on Pacific Lumber property, with one exception: with due notice to the company, I would be allowed on the preserved area around Luna.

We had to agree to establish guidelines should either one of us break our word. I opted for mediation, with arbitration as a fallback should that not work. I wanted to keep as close as possible to the spirit of people interacting with each other rather than automatically turning a conflict over to the legal system.

Next came a complicated time line of schedules and procedures involving removal of the fallen timber. An initial survey would be followed by an on-site meeting between foresters from each side to determine what could be taken and how.

IN THE MIDST of all the legal back-and-forth, Redwood Mary, a talkative East Coaster whose involvement with the forest dates back many years, called me. An incredible woman named Wangari, who had been working in Kenya to preserve what little is left of the forest there and restore what's been damaged, was going to hold a huge action for the African greenbelt movement on July 3, 1999.

"We need to think of something to do in solidarity," Redwood Mary said.

We began to brainstorm over the phone. The first thing that came to mind was an Independence Day celebration for the forest—independence from the greed and consumerism that are destroying our life support system. But after sitting on the idea for a bit, I called her back.

"I have a better idea," I said. "Let's try Interdependence Day instead. Independence from greed and overconsumption is more about nagging people. Interdependence Day is about celebrating the interconnected web of life. We need to create something that's positive, not negative."

So we celebrated July 3 with Interdependence Day, to help people understand how vitally and intricately interconnected—and therefore interdependent—all of life is, from the air that we breathe to the largest redwood trees to the little microscopic bugs and fungi. It's one life, and we are in the middle of it all, dependent on all the species alive whether we know it or not. Every time a species goes extinct, a part of our life has been altered forever. You cannot rip out one thread and not have the whole tapestry begin to unravel.

The idea caught on. In short order, people started holding prayer circles and music circles and tree plantings in New York, New Jersey, Missouri, and here in Humboldt and Mendocino counties. It also took off in Israel, the Philippines, and Germany, where activists want us to come and kick off next year's observance. For something we threw together in two weeks, we did okay.

NEGOTIATIONS WITH Pacific Lumber proceeded at the opposite pace. Finally, on July 14, 1999, a historic meeting took place between Maxxam's lawyers, Dale Head and Eric Erickson, a representative from the title company, and the two people from my team, Tryphena Lewis, my trusted friend and former Trees Foundation support coordinator, and Herb Schwartz, a mediation lawyer. After reviewing the various points of the tediously crafted, legally binding agreement, including the time line to make it happen, the parties realized that there no longer remained any major disagreements. Our side asked for a word that Maxxam had thrown in at the last minute to be taken out. (It was a vague word that would have required four pages to explain.) They agreed, and that was it. The last bone of contention had been resolved.

The written document embodied the spirit of verbal agreement that John Campbell and I had reached. It didn't bind the company to anything it had not agreed to, but it protected Luna and her surroundings. A true resolution, where both parties felt they had won. All that remained was for the surveys to be done and for the agreement to be retyped, signed, and delivered to the title company. In a matter of weeks, it would be over.

I knew that the mainstream television media would want to talk with me when I finally came down from Luna, so I started scheduling interviews. One was for *The Late Show with David Letterman,* of all places! It was wild to imagine going from sitting in a tree to being a guest on a show that so many people I knew watched. I think the segment producer, however, was concerned that I might prove too serious a guest. She called to preinterview me.

"I'm going to have to ask you some questions," she cautioned.

"Like where do I go to the bathroom, how do I shower, and do I have a boyfriend?" I interjected.

"How did you know?" she asked in amazement.

"Well, I go to the bathroom like everyone else, but I use a bucket. I take sponge bathes using water that I collect in my tarp. And who needs a boyfriend? I have a tree." The segment producer was quiet for a second and then completely broke up.

So much for being too serious.

⌒⌒⌒

STILL, DESPITE ALL the preparations, I had to protect myself against wanting to put my feet back on the ground. Yes, if I had allowed myself to, I would have felt excited about going down and looking at the world I had left through new eyes. I would have felt excited about indulging in those creature comforts I'd left behind over eighteen months earlier. But I knew that holding expectations only invites disappointment. So I refused to get my hopes up.

The people who worked with me on the deal that would protect Luna and the grove couldn't understand why I wasn't more excited. During the negotiation, they jumped at every deadline and tried to make it fit into a time line. When that time line passed, they would get all worked up over a new one. If I had operated that way, I could never have lasted over nineteen months in a tree. I would have burned out a long time ago.

Things were spinning really fast. If I got caught in the spinning, I would lose sight of my focus. I just had to take the appropriate step as it unfolded, exactly as I had done almost two years before. I did not sit down on the ground and plan this whole tree-sit. One step led to another, which led up the hill and into Luna.

Yes, after nineteen months I wanted to come down. I knew it might happen soon. And I had to leave it at that.

Good thing, because instead of signing, Pacific Lumber stalled. I called Campbell.

"How do we get this moving forward?" I asked.

"I'm really swamped right now," he responded. "I'm trying to get a couple more people to agree to this. They're worried about what you will say to the press when you come down.

But trust me, I'm working on it. I'm building a constituency."

He promised to call me the next day. He didn't.

Someone had apparently put the brakes on our agreement, for whatever reason. I couldn't get a straight answer. Maybe Hurwitz himself had stepped in. Or maybe they had just changed their minds about whether there was really enough in it for them.

If they were worried about what I was going to say to the press, they should have protected the area as quickly as possible. For as long as I was up in Luna, I would slam home destructive forest practices every single day. During our negotiations, I had actually tried to stall people from staging a number of big demonstrations designed to promote protection of the area. Though these would have helped spread the word, I felt that any demonstration would have been a breach of what I believed to be mutual good faith. Obviously, I had more faith in Pacific Lumber; me and my eternal optimism. Still, I held the hope that we might resurrect our pact.

"John Campbell wants to do the right thing," I told myself. "He knows that protecting Luna and this grove stretches far beyond the two of us. He'll uphold his part of the deal. He acknowledged in part of the agreement Luna's public interest and historical value."

In the next breath, I berated myself for having expectations. I knew better.

Finally, I went back to just taking it day by day. I still had an incredible amount of work to do.

"I can sit up here for the next three years and keep working my butt off," I thought. "If they want to give me free rent while I continue to hammer them every chance I get, so be it. I'll continue to make them work for their $4,000,000-a-year public relations firm. They'll need it more than ever, because I will not slow down."

I wanted to protect Luna for her sake, for the sake of the hillside, and for the sake of the people in Stafford, whose voices I heard quivering as they described what it was like going to bed at night knowing that the hillside might at any minute give way and bury them in mud. I wanted to protect Luna for the thousands of people across the country and around the world for whom she had become a symbol of hope, a reminder that we can find peaceful, loving ways to solve our conflicts and that we can take care of our needs without destroying those needs to satisfy our greed.

It seemed so simple from where I sat, because I could see mud slides, and every one of them began on a logging road in a clear-cut. I saw miles and miles of clear-cuts, where the burned, desecrated land struggled to hold itself up because the forest that should have been supporting it had been destroyed. I watched as the creek beneath me, brown with sediment, poured into the Eel River, knowing that clear-cuts line the creek for miles on the steep slopes above. As I watched all that nutrient-rich soil, which will take hundreds and hundreds of years to replace, being washed away, I mourned for the ecosystem that we were losing in that process, this healthy soil needed for the plants that provide us oxygen, food, and temperate climes. We humans are not the creators of life, but it seemed to me that we were on the fast track to being its destroyers.

So I stayed up, and life in Luna, with all its peculiarities and incongruities, went on as before. One night in August, my regular phone again lost its signal. I picked up my cell phone, my only remaining form of communication, but it suddenly cut out while I was mid-sentence. Then a stranger's voice came on.

"Hey, what's going on?" I exclaimed. "Who are you?"

"I'm from the phone company," she answered abruptly. "What's the number of your phone?"

I gave it to her.

"What is the name on your account?"

"I'm not really sure at this point. It's either Robert Parker, Julia Butterfly, Julia Butterfly Hill, or Josh Brown." (He's the Earth First! organizer who allocates resources for the campaign.)

"Oh, yes, it's under Josh Brown, attention Robert Parker. Is either one of them around?"

By this time I was beginning to suspect that somehow the phone bill hadn't made it into the hands of the right people and had never gotten paid.

"I need to speak to one of them, because they're the only ones on the list," she continued.

"Okay, let me try to explain," I said. "I'm living in a tree as a form of protest against the cutting down of the ancient redwoods. Josh Brown was the first person who had this cell phone under his name. After that we switched it to Robert Parker. We've been paying the bill whenever we get it."

"Well, I can't do anything about all that. I'm going to have to cut off your phone."

"Wait!" I almost yelled. "My normal line of communication is cut off. My very life depends on this cell phone right now."

She hesitated. I could feel her trying to decide whether to follow the book or to have a little heart for this woman with the bizarre story.

"I'm sorry," she repeated after a few seconds. "I need to talk to . . . "

I played the only card I had left.

"But I live in a tree, and the only people who can pay the bill have to hear from me to learn that they have to pay it. You have to turn my phone back on, or you'll never get your money."

I don't know whether she ever really understood what was going on, but she finally agreed.

"I'll turn it back on, but only until tomorrow afternoon," she conceded. "If you don't get it paid by tomorrow afternoon, we're cutting it off."

What next?

So much for the myth of the woman who lives alone in a tree, her life all peaceful and idyllic.

⸻

PACIFIC LUMBER'S RIGHT to cut on the land around Luna expires in September 2000. Of course, the company could probably renew it without a hitch, despite scientists attacks over their "environmental strategies" and despite their pleas of "guilty" to criminal violations of the Forest Practices Act and despite all their consistent violations. On the other hand, if I'm still around, they may not want to.

One thing I am constantly reminded of is that despite all the plotting and strategizing, things never work out quite the way anybody intended. But whatever happens in the future, my stay in the tree did answer my prayer that day on the Lost Coast. Luna changed me. Living in this tree, I remembered how to listen, to hear the world and Creation speak to me. I remembered how to feel the connection and conscious oneness that's buried deep inside each of us.

So I will continue to stand for what I believe in, and I will continue to refuse to back down and go away. No person, no business, and no government has the right to destroy the gift of life. No one has the right to steal from the future in order to make a quick buck today. Enough is enough. It's time we as humans return to living only off the Earth's interest instead of drawing from the principal. And it's time we restored some of the capital investment that we've already stolen.

It is our responsibility to stand up for the life we've recklessly squandered, no matter what the consequences. So I'll continue to hold the light strong even in the midst of total darkness. I will continue to believe that love is the answer, love is the power, love is the truth.

Creation did not mess up. It didn't go, "Oops, I didn't mean that to happen." When you rip a plant from the roots that connect it to the soil and life, it dies. So when we rip out the roots that connect us to Creation's love, we also die.

So I will continue to hold Creation's love in my heart, in my actions, in my truth, my words, and my very thoughts. I am human, very human, so I will stumble, and I may even fall. But I will pick myself back up again and hold on to love for all I'm worth and try again.

Luna is only one tree. We will save her, but we will lose others. The more we stand up and demand change, though, the more things will improve. I ask myself sometimes whether the destruction has gone too far, whether we can really do anything to save our forests and our planet. And yet I know that I can't give up. We must do the right thing because it is the right thing to do regardless of the outcome. I have to take it one struggle at a time. And just as I've done with Luna, when that struggle comes my way, I've got to fight it for all I'm worth.

Yes, one person *can* make a difference. Each one of us does.

EPILOGUE

I was settling in for another winter feeling that—for whatever reason–the people at Pacific Lumber/Maxxam had decided not to go through with their end of the agreement to protect Luna permanently. Like a squirrel gathering up its crumbs and stashing them in its hiding places, I began gathering clothes and supplies and stashing them in the nooks and crannies of my home. I was feeling a little worn out knowing I was facing a third winter in Luna. I trusted that the Creator would give me the strength and the clarity to keep going, but I was also feeling very human. I wished for some comfort, and I had a slight sense of fear, too. I knew that the past two winters I spent in the tree had put my life at the very edge. The thought of facing yet another made me feel even closer to whatever lay on the other side of that edge. I knew that no matter what was going to happen, it was going to require a lot of prayer on my part to keep going.

I didn't know what the future would hold, when, out of the blue, a representative from Maxxam responded to a call from

John Goodman, a striking Kaiser Aluminum steelworker who had been calling Maxxam on my behalf for many months. It will always remain a mystery to me that one of their own disaffected workers would be the one to help break a seemingly impossible stalemate. Goodman began calling Maxxam on my behalf after hearing David Chain's mother speak to Charles Hurwitz at a Maxxam shareholders meeting. At that meeting, she had stood up and faced Hurwitz. She told him that she had lost a son on Maxxam property, and asked what he was going to do to make sure that no mother ever received a telephone call, as she had, saying that her son had been killed. In response, Hurwitz promised that he would do whatever he could to ensure that an activist would not be hurt or killed on his land again.

John Goodman knew of my commitment, and he was very concerned that I, too, could die if an agreement was not reached. Goodman finally got in touch with Jared Carter, the acting vice president and in-house counsel for Pacific Lumber. They began talks on finding new resolution to bring safe closure to the tree-sit.

Negotiations resumed between Goodman and Carter. Goodman acted on my behalf since Carter wouldn't talk to me directly. From the beginning, the talks were conducted in what I have come to call the bully-bluff system. I had studied Hurwitz for two years by this time. I knew how he negotiated. I knew what he was doing. He did it when he purchased Pacific Lumber. He did it with Headwaters. And he had done it with me. What he does is push to the limit and then back out. By doing that, he gets a lot of concessions.

I was raised playing chess, and Charles Hurwitz is one master chess player. He is so good that if he were only in touch with his heart, he could be one of the major forces for good in the world. But, having been matched against him before, I knew his strategies. And I was prepared. He was going to try to move me around like a pawn. But I stayed focused on my goal, which was for me a resolution where all sides could win. This was not a competition between Julia Butterfly and Charles Hurwitz; this was about the whole world winning.

I knew I would have to make more concessions to empower the company to do the right thing. But this time, they started placing demands that would limit, restrict, and control my right to free speech. They were trying to get me to sign statements that would basically discredit direct action, tree-sitting, and the forest activist movement, while having me pat them on the back for their Habitat Conservation Plan, which I believe is bad for endangered species, our planet, and our lives. This was a new move: they were holding my beliefs and desires hostage. I drew the line, however, and refused to sign away my values, my morals, my beliefs, and my rights.

What I was willing to do, however, was listen to anything else they proposed, even when they were making extremely unreasonable demands. Almost the full duration of the negotiations revolved around trying to restrict what I would say and do with my life after I came down. I tried to let down my wall of anger and frustration in order to hear what was under the really ridiculous things they were demanding and that took a long time.

Up to this point, Pacific Lumber/Maxxam and I had honored an informal confidentiality agreement that was understood between us, which allowed us space to negotiate. But John Campbell began leaking information about our talks to the public and to the press. Campbell was the first one, but then a Maxxam spokesperson called me a liar in the press. He said that I was the one who had first leaked the details of our discussions, but I had recorded, documented proof that I was the one telling the truth.

Now that our discussions had entered a public forum, I finally felt that I had to call a press conference to set some things straight. I told the press that they had an obligation as journalists to present both sides of the story so their audiences could make up their own minds about what was going on.

I had a chance to speak my truth, and the public rallied around me. It wasn't very long until that support was brought to bear on Senator Dianne Feinstein who, in turn, pressured Pacific Lumber/Maxxam to back off of their unreasonable restricting demands. What I believe happened was that final pressure compounded every single action of the past two years—no matter how seemingly insignificant the act—and finally tipped the scales. Pacific Lumber/Maxxam was ready to do the right thing.

Shortly after the press conference, the resolution began to

take shape. Tryphena Lewis, a member of the Circle of Life Foundation team, joined the discussions. She made a critical difference as she provided an element that had been lacking. Unlike Carter and Goodman, she had been a part of the first round of negotiations. She kept everyone centered on the intent of the first agreement, abandoned by Pacific Lumber, which would save Luna and create a buffer zone around her. With behind-the-scenes help from attorneys Sharon Duggan, Herb Schwartz, and Tom Ballanco, the final points were worked through.

On December 18, 1999, a preservation agreement and deed of covenant to protect Luna and create a 200-foot buffer zone into perpetuity was documented and recorded.

I had kept up with the negotiations, but hadn't allowed myself to feel too hopeful. After all, things had fallen apart at the last minute before. So when the title officer called me on a speaker phone and said in a serious voice, "I have something to tell you," the first thought that ran through my head was, "Oh, they found something in the report that nullifies all this hard work!" But in the background, I could hear Tryphena, and there was something in her voice that betrayed her excitement. Then he said, "I have really good news for you. The document has been recorded. The Luna Preservation Agreement and Deed of Covenant is done."

"I think really good news is the understatement of the millennium!" I said but still didn't feel much of anything. I think I was numb. But the moment I hung up the telephone, it hit

December 18, 1999.

me. I had been standing up, and I fell to the platform and cried. It was finally done. No more loopholes. No more stalls. Luna was protected. We did it.

Throughout the entire negotiation process, I hadn't allowed myself to feel anything. I had to act without attachment to what was going on. Otherwise, I would get too worn out riding my feelings up and down. But it was safe now. And the emotions hit me like a tidal wave.

What was wonderful was that when I received the news, my main support team was gathered at the base of Luna. I had asked Michael, my ground support coordinator, to bring some people up with him so that they could begin packing up things I had accumulated over two years. I was planning to hold on to my winter gear, but it was still time to clean out. So Michael, Spruce, Shunka (my dear friend from many years ago), and K.C. (who had been apart of the first group who had assembled the first platform in Luna) were all there when I got the news, and it was phenomenal.

At last, it was time to go. During one of my last nightly climbs to the top platform to plug in a battery to the charging system, I climbed out onto one of my favorite branches of Luna and nestled in to her loving embrace. It was a gorgeous evening with the fog slowly rolling into the Eel River valley down below and a crystal clear sky above. The waxing moon brightly lit much of the sky and made the fog glow iridescently. A few stars shimmered at the edge of my view. It was a scene I had seen many times over the past two years—each time slightly different, each time breathtaking. As I sat there taking it all in, I realized that this was to be one of the last times I would ever look out at this view that had become as much a part of me as the experience had. I burst into tears.

Later that night I wrote, "I feel like I'm being separated from a part of myself—a piece of me—the essence of who I am. The woman I have become is being torn right now. I am beginning to feel the understanding—the never-ending lesson of letting go. When I leave this tree, I will be leaving the best friend I've ever had. It is a pain I can not describe, only feel. . . and be with. I am with it now, and it is only the beginning. I will be able to come back and be with Luna in her womb at her base, but never again will I perch in her branches viewing the world from this incredible perspective. I will do my best to live the rest of my life in honor of her and this experience—offering myself as the only gift I have to give. It is my prayer and my hope that I will always and only be an offering."

A few days later, as I prepared to descend Luna for the first time in 738 days, which would also be the last time, I began to sob once again. How would I be able to keep the focus, grounding, and truth that I had found in Luna? How would I be able to keep going when it felt like I was dying, having to leave this incredible living being? I prayed, and for the last time in her branches, Luna spoke to me and reminded me of something I had received in prayer nearly a year before: "Julia, all you have to do when you are afraid, lonely, worn out, or overwhelmed is touch your heart. Because it is there that I truly am, and it is there I will always be."

Julia Butterfly Hill

December 23, 1999

AFTERWORD

The Sunday afternoon after Thanksgiving had been a peaceful one in Florida, where I was visiting my grandparents. It was a long overdue visit, one I was reluctant to interrupt with work, but I knew I had a phone interview coming up that evening so I paged my manager, Paul Bassis—known to all as PB—to find out when it was. Moments later, the phone rang.

"Hi, PB." I answered. "I was calling to get the details on that phone interview tonight."

"Okay. Sure," he seemed to stumble. "Hi . . . Julia . . . how are you?"

Immediately, I could tell that something was wrong. "I'm fine PB," I told him firmly. "What's wrong?"

"Are you really okay?" he asked.

"Yes. I'm fine. Why? What's going on PB? What's the matter?"

"Julia, I don't know how to tell you this, but . . ." his voice trailed off.

"PB, tell me. What's going on?"

"Julia . . . Luna's been cut."

"What?"

"Luna's been cut, Julia."

I went numb. I felt as if I had been kicked in my stomach and all the air sucked out of me.

"Julia?"

"Yeah PB. I'm here. How bad is it? Is she down?"

"No. She's not down. But it's not good Julia."

"Who found her?"

"Shunka. I'm waiting for a call from Claudia. She's gone up to see how bad it is, but Shunka said it's most of the way through. I was waiting to call you until I had heard back from Claudia."

In the background I could hear PB's pager going off.

"That's Claudia now, Julia."

"Okay."

"Julia?"

"Yeah?"

"I'm so sorry. I'm so sorry to be the one having to tell you this . . . I'm just so sorry."

"It's okay, PB. I thank you for being the one to tell me. I know it's not easy, and I truly appreciate you for being the one."

"Are you going to be okay, Julia?"

"Yeah, PB. I'll be fine."

"Sure?"

"Yeah. Thanks."

But I wasn't fine. I wasn't going to be okay. But now was not the time for me to admit that. Not to PB. Not to myself.

"Okay. I'll call you back as soon as I know more."

"Alright. Thanks."

I felt like I was outside of my body seeing myself from a distance. I could hear my flat tone speaking calmly as if I hadn't just been given the worst news I could ever have heard. I was very practical and information oriented. I was acting as if this vicious attack on Luna was nothing more than a detail to be worked out. I guess I was in shock. I *know* I was in shock. As soon as I hung up the phone with Paul, it felt as if all the air that had been knocked out of me came whooshing back in, hitting me all at once and I was shattering underneath the force. I ran out into my grandparent's front yard because it was the closest to nature I could get at that moment. I sank my face and fingers into the ground and all I could do was wail, "No, No, No," a moaning mantra that threatened to rip me apart.

I knew my grandparents would be coming to see what I was doing outside, so I hurriedly built a wall around my pain, wiped the tears from my eyes, and headed inside.

"What's wrong, Julia?" my grandmother asked me. So much for my defenses.

"Well . . . um . . . some . . . someone attacked Luna with a chainsaw. They cut through about two-thirds of her from what it seems. They don't know if she is going to be able to survive," I said choking with tears.

I was dying inside. But the need to maintain a clear head helped keep the pain distant enough that I was able to make the necessary arrangements to fly back to California and I boarded the first plane home. By evening, I was back in Humbolt county, lying on a sofa in the Circle of Life Foundation office, my attempt at sleep fitful and troubled. It was one of the longest nights of my life as I waited to join a group that would go see Luna the next morning.

I dreaded the moment when I would first see Luna because I knew that as a public figure, I would have to go through the grief that was waiting to engulf me in front of cameras, microphones, and the press. From a personal point of view, I wanted to beg the world to go away and let me confront this tragedy by myself. But the life of service to which I had surrendered while living in Luna's branches compelled me to be willing to feel the overwhelming sadness in front of everyone. While I wanted privacy, I knew, too, that part of my path meant showing my feelings so that others, too, could grieve, learn, and hopefully heal.

The drive to Stafford was both an eternity and a fleeting moment. From the highway, I could see Luna standing tall at the mountaintop. From that great distance, it seemed that nothing was wrong; that she was fine. How I wished that were so. We slowly drove up the rutted mountain road led by an orange Pacific Lumber truck passing new and horrific clearcuts. Huge, fresh gashes were cut into the steep hillsides. Many of them had been burned, the earth blackened and completely devastated and desecrated.

The irony made me sick to my stomach. Here, a representatives of Pacific Lumber were leading us to the injured Luna, while their co-workers were out slaughtering hundreds if not thousands of other trees just as special as Luna—annihilating whole forest ecosystems within a matter of days.

We reached the line where cars must park, and the hike begins. Out jumped all the people. Out jumped the film crews, radio person, and photographer. Out I climbed. Out into a moment that I was praying was just a bad dream from which I would awaken.

At different points along the trail, I stopped and could see Luna standing high above the trees around her. I felt some of the same profound waves of emotions washing over me that I had felt when I walked down the same path on December 18, 1999, and when I returned for the first time in June of 2000. There was joy that Luna still stood but a simultaneous deep sadness. But there was not much time for reflection. I felt the cameras zooming in on the tears that were beginning to fall, pushing their way past the barriers I thought I'd constructed to keep myself "safe" in public. I was reminded of the poem I had written the night before my first return to Luna,

The world stops
and holds its breath
and wonders
What will "she" do?
What will "she" say?
What does "she" see?
As if "she" is something other than
me . . .

As I scrambled my way up the last, extremely steep, slippery, pebble path, every step I took forward threatened to send me three feet back. I reached the crest, and there Luna stood in all her beauty and power. From there, before a short walk down to get to her base, all I could see was the strong Luna I knew and remembered.

Luna's energy and that of what is left of the grove around her hit me very strongly. I felt the weight of what I was about to witness before I even saw the damage. And as I dipped down to eye-level to Luna's base, I saw it.

The cut. The deep wound screeched and howled right through me. The metal braces that had been placed on her to stabilize her so she wouldn't snap in a high wind. Big. Scary. Awful. I couldn't help myself; I collapsed onto the ground before I even reached her base. No was the only word to escape through my grief and tears. Over and over, louder and louder. "No, no, nooooo."

I wanted to scream but I became aware of the cameras closely trained on my face, so I checked the pain. Standing, I hesitantly walked the remainder of the space between my dear friend and me. My hand slowly tracing its way around the deep wound, "I am so sorry. I am so sorry. I am so sorry," was all I could seem to say. How do you tell your best friend in the Intensive Care Unit in the hospital after an attack that you wish that you could have been there for them to stop their attacker? All I could think of was how much I wished that chainsaw had gone through me instead.

I told everyone I was going to pray and encouraged them to do their own form of prayers if they wanted. I lit sage, asking that we be cleansed. I prayed for healing for Luna and for the world. I prayed for the person or persons who so viciously attacked her. I acknowledged their pain and rage. How else could they be so motivated to attack something that can not defend itself; that cannot run away. Whoever performed this horrid act needed healing as well, for they must be riddled with profound hatred and anger for the world.

After the prayers and spending time with Luna, it was time to head back down the hill to a press conference in Eureka. Now more than ever, I needed the lessons I learned living in her branches. Bend. Flow. Let it go. Let go of the grief. Let go of the human attachment to want to be by myself. Go talk to the media. And pray and hope that they get it and do the real story. The kind of story they should be covering. To see that the Luna action always has and always will be about more than one tree and one person.

Luna stands as a symbol. A symbol for all the old-growth forests that are smashing into the ground, into oblivion, into extinction, every day. Luna stands for hope and the love that will always win over hate. She reminds us that there are no "sides", only "us"; that love and hate are within us all. Luna reminds us that the hope for this beautiful, sacred planet that gives us life and thus hope for our humanity lies in our ability to transform the greatest obstacles and challenges into strength, endurance, commitment, and love. These are the essence of Luna.

So I did the press conference. I gave it my best. I gave it my all.

Some of the coverage was good. Some of the coverage was shallow, typical hype. Some of the coverage and resulting backlash was mean, petty, and hateful. But as I learned a long time ago, media will be media, and all I can do is continue to do my best with every situation.

There will be times of blissful joy beyond my wildest imagination and times of pain where I truly wish that I could die. There will always be points along my path that will challenge me with struggles and adversities that seem at first glance too big to take on. Too big to change. Too big to transform. But as I learned in Luna's branches, in the most powerful, sacred time of my life, the difference is in the doing. We do the right thing, because it is the right thing to do, regardless of the outcome. We persevere because someone today has to hold those accountable for the children of tomorrow. We love with all that we have, even in the face of overwhelming hatred and violence because we know love in action is the hope for our world.

Yes, symbols can be attacked, but what they stand for can never be destroyed. The assault on Luna is an assault on what so many in this world hold sacred and dear. But instead of weakening or destroying us, it has only served to make our resolve stronger. Whenever Luna's time comes to fall into the forest floor, she will be giving herself to feed new life that will grow from her gift just as she grew from one before. Her nutrients will be a part of an ecosystem that proves that a forest is more than just a tree farm. It is a relationship, symbiotic, important beautiful. The same with our planet and our lives. Yes, the wound in Luna and our world is deep. But so is our commitment to healing.

Julia Butterfly Hill

January 1, 2001

ACKNOWLEDGMENTS

In the beginning, there was Love, and Love created and thus became Creator. I give all honor and glory and my humblest gratitude to the Creator for giving me the gift of Life in Love held so delicately in each and every breath, and for guiding me to an experience where I could learn the power, magic, and understanding of just how incredible this gift truly is. For the rest of my life, I will do everything within my power to honor, cherish, and protect this gift of life that connects us all.

From the very depths of my being, I give my heartfelt gratitude to each and every person that has ever helped me in my path in life, and especially everyone who helped this action in any way. To each person who donated support on any level, who shared prayer and encouragement, who helped spread the word, and who helped on any and every level, I say thank you in a way I will never be able to describe fully.

My humblest gratitude to everyone who has ever risked arrest, past, present, or future; been arrested; or is currently imprisoned for standing up for their beliefs. My prayers and heart are with you. At the time of this writing, there is a person very important, courageous, and powerful who is locked up for being so and who needs your help. His name is Leonard Peltier. Leonard Peltier has been wrongfully imprisoned for nearly 25 years for a murder he never committed. The government knows he is not guilty; there is overwhelming evidence exonerating him, even members of the prosecution say they do not know who committed the murder that Leonard is accused of, yet he, like many other political prisoners, remains locked up as a scapegoat for our government's stupidity and violence. Leonard's imprisonment is all of our shame. He and every native person needs our help in ending America's long history of genocide against indigenous peoples. It is up to each and every one of us to seek justice and freedom for this phenomenal human being. Please see the resources page for contact information and get involved in helping Leonard find freedom and justice.

My deepest heartfelt gratitude and love to everyone. We did it! There is much more to be done, and I thank everyone who will continue to do their part. Everyone who faces fear, oppression, and violence yet remains rooted in love, commitment, and action is my hero.

In service of life in love,
Julia Butterfly Hill

RESOURCES

Circle of Life Foundation
P.O. Box 388
Garberville, CA 95542
(707) 923–9522
www.circleoflifefoundation.org

Ancient Forest International
P.O. Box 1850
Redway, CA 95560
(707) 923–3015
www.ancientforests.org

Bay Area Coalition for Headwaters (BACH)
2530 San Pablo Avenue
Berkeley, CA 94702
(510) 548–3113

Environmental Protection Information Center (EPIC)
P.O. Box 397
Garberville, CA 95542
(707) 923–2931
www.wildcalifornia.org

Environmentally Sound Promotions (ESP)
P.O. Box 2254
Redway, CA 95560
(707) 923–4646
www.jailhurwitz.com

Free Mumia Abu-Jamal Coalition
International Concerned Family and Friends of
    Mumia Abu-Jamal
P.O. Box 19079
Philadelphia, PA 19143
(215) 476–8812
www.mumia.org

Friends of the Eel River
Headwaters Action Video Collective
P.O. Box 1655
Redway, CA 95560
(707) 923–0012

Institute for Sustainable Forestry
P.O. Box 1580
Redway, CA 95560
(707) 247–1101

Leonard Peltier Defense Committee
P.O. Box 583
Lawrence, KS 66044
www.freepeltier.org

Mendocino Environmental Center
106 W. Standley
Ukiah, CA 95482
(707) 468–1660

Native American Coalition for Headwaters
They can be contacted through the Trees Foundation.

North Coast Earth First!
P.O. Box 28
Arcata, CA 95518
(707) 825–6598

Pepperspray Legal Fund
They can be contacted through the Trees Foundation.

Rainforest Action Network
221 Pine Street
San Francisco, CA 94104
(415) 398–4404
www.ran.org

Ruckus Society
2054 University Avenue, #204
Berkeley, CA 94704
(510) 848–9565
www.ruckus.org

Sanctuary Forest
P.O. Box 166
Whitethorn, CA 95589
(707) 986–1087

Trees Foundation
P.O. Box 2202
Redway, CA 95560
(707) 923–4377
www.treesfoundation.org